Energy in Africa

Sola Adesola • Feargal Brennan
Editors

Energy in Africa

Policy, Management
and Sustainability

Editors
Sola Adesola
Oxford Brookes Business School
Oxford Brookes University
Oxford, UK

Feargal Brennan
University of Strathclyde
Glasgow, UK

ISBN 978-3-319-91300-1   ISBN 978-3-319-91301-8   (eBook)
https://doi.org/10.1007/978-3-319-91301-8

Library of Congress Control Number: 2018953179

This Palgrave Macmillan imprint is published by the registered company Springer Nature Switzerland AG
The registered company address is: Gewerbestrasse 11, 6330 Cham, Switzerland

The original version of this chapter was revised. Footnotes, citations and references were added. Correction to this chapter can be found at https://doi.org/10.1007/978-3-319-91301-8_10

Preface

The Origin of This Book

This book was originally inspired by the first author's senior academic colleague, Professor Glauco De Vita, with whom they both successfully supervised two PhD researchers to completion on local content policy and stakeholder analysis, and energy demands in Sub-Saharan Africa at Oxford Brookes University. Just after Professor De Vita left, he sowed a seed of encouragement into Sola Adesola to consider publishing a book; at that time, energy was never on the plan. The idea of a book then was on entrepreneurship in Africa. From mid-2016, however, Dr Adesola had been asked to be an international panel expert for the Nigeria Energy Forum, and a panel member at the Nigerian Society Energy Conference at Cranfield University. She also had the opportunity to peer review energy conference papers. Prior to the conference, she had an unforgettable opportunity to share her interest in energy from past voluntary reviews of Cranfield PhD thesis on energy-related research and mock vivas for Cranfield students. This strategic meeting was with Professor Feargal Brennan, then the Director of Energy and Power at Cranfield University, now at the University of Strathclyde, who later became the co-editor of this book. It was at this meeting that the thought of publishing a book on energy was conceived. The first author left that meeting with the golden opportunity to present a "call for book chapters" at the

2016 Energy Conference in Cranfield. At this stage, no publisher had been contacted, but we proceeded nevertheless.

This book is the result of nearly two years' intensive work. It started by first recognising the gaps in energy research that exist between engineering, commerce, socio-politics, environment, and management. Following a quick Internet search, it became apparent that what was needed was a collection of sound academic research and policy papers on energy from a management perspective focusing on this important continent—Africa. This edited book is a first attempt to collate interdisciplinary chapters in one book volume. It is based, in part, on our collective involvement over the past 25 years, in energy and power by Professor Feargal Brennan as an internationally renowned expert, and on international business and African Studies by Dr Sola Adesola.

The second reason why we felt that a new and different sort of book on energy in Africa was needed is due to the changes taking place in the wider world of international business and international organisations which are altering the landscape of Africa. This book attempts to bridge the language barrier between technical and soft issues. Feargal Brennan makes the related point that energy is an international sector and most of the largest and most exciting infrastructural developments ever seen are poised to take place in the developing world over the coming decades. Africa is a wonderful example and case study of the multifaceted socio-economic-political-environmental criteria which influence the adoption of technologies, balancing local content within an international context with lessons and learning that can be applied to many other developing regions.

The third reason why we thought that a different sort of book on energy in Africa was needed is based more directly on the first author's own experiences and encounters in research on energy for more than eight years. To explain this, between 2008 and 2017, Sola Adesola was fortunate to be the co-pioneer of the PhD student mentoring support as part of the local church outreach at Cranfield University. She participated in voluntary reviewing of thesis and organising mock vivas for a number of Cranfield students. She became an "adopted" engineer from familiarisation with engineering technical jargon. Most times she found herself asking questions relating to the managerial or policy implications of their research, and most times, these had to be extracted from facilitation.

Over time, the authors have undertaken a succession of internationally recognised peer-reviewed funded large research projects, often in collaboration with others; other times the authors have incorporated energy case studies into the teaching curricula for coursework assessment and presentation. The outcomes of years of engagement in energy research and scholarly work have resulted in world-class publications.

This book is compiled, in part, as a reference academic book, to help those engaged in energy engineering, policy, and management in Africa to better understand how they "fit in" together.

Oxford Brookes Business School   Sola Adesola
Oxford Brookes University
Oxford, UK
University of Strathclyde   Feargal Brennan
Glasgow, UK

Acknowledgements

This book volume is a snapshot of a continuing conversation about energy in Africa that we have been fortunate enough to have had with colleagues, students, and friends. From the early days, we owe a particular appreciation to Professor Glauco De Vita at Coventry University, who encouraged the first author to undertake book chapter publishing, and who, to this day, remains one of the most renowned scholars on international business economics. We are particularly grateful for his unbiased and constructive review and for his helpful suggestions for improving its structure and contents. Most immediately, this book would not have been completed without the immense and unrivalled support and collaboration of the co-editor, Professor Feargal Brennan and Cranfield University Energy Conference platform for the initial introduction to get the idea started.

Many academics and international *experts* have been generous in giving their time to critically *review the chapters* of the book as they have emerged. Their comments and suggestions were of great value. Among those who have provided assistance and support, we would like, in particular, to thank the following:

Professor Glauco De Vita   Coventry University, UK
Professor Damilola Olawuyi   Hamad Bin Khalifa University, Qatar

Professor Chuks Okereke	University of Reading, UK
Professor Leo Daniel	MIT, USA
Dr Uyiogbosa Igie	Cranfield University, UK
Dr Simon Adderley	Oxford Brookes University, UK
Dr Georgina Whyatt	Oxford Brookes University, UK
Maureen McCulloch	Oxford Brookes University, UK
Dr Elias Boukrami	Regents University London, UK

The individuals that reviewed the chapters to this book are not responsible for any opinions or judgements it contains. All errors and omissions are solely the responsibility of the authors.

We would like to thank our copy editor, Mrs Heather Simpkins, for her editorial support for the correction of the chapters, her eye for detailed comments and suggstions for improving language and presentation.

Most importantly, a book like this could not have been completed without the contributions of the authors of the chapters, who have shared their research and practice with the world, and have raised critical issues of policy, management, and sustainability from African energy perspectives. We thank and acknowledge your contributions to Africa.

Finally, we thank Palgrave publishers who have helped to steer the book to publication. We are also grateful to Lucy Kidwell for her invaluable assistance and patience. It has been a major undertaking simultaneously during academic year.

Contents

Notes on Contributors

Sola Adesola is Senior Lecturer in International Organisations and Corporate & Business Law in the Oxford Brookes Business School at Oxford Brookes University, UK, having previously completed her PhD and held research position in manufacturing management at Cranfield University. Adesola is Director of Studies for research in energy policy, entrepreneurship, and university interaction. She has written widely in peer-reviewed journals and conferences including the following publications:

Kolawole, A., S. Adesola, and De Vita. 2017. Factors that drive energy use in Africa: Panel Data evidence from selected SSA Countries. *OPEC Energy Review* Dec, 41 (4). doi:https://doi.org/10.1111/opec.12115.

Onabanjo, T., S. Adesola, and A. Kolawole. 2017. A state-level assessment of the electricity generation potential of municipal solid wastes in Nigeria. *Journal of Cleaner Production* doi:https://doi.org/10.1016/j.jclepro.2017.06.228.

Kolawole, A., S. Adesola, and G. De Vita. 2017. A Disaggregated Analysis of Energy Demand in Sub-Saharan Africa. *International Journal of Energy Economics and Policy* 7 (2): 224–235.

De Vita, G., O. Lagoke, and S. Adesola. 2015. Nigeria oil and gas industry local content development: A stakeholder analysis. *Public Policy & Administration* 31 (1): 51–79, doi:https://doi.org/10.1177/0952076715581635.

Kenneth E. Afe Aidelojie was former Senior Lecturer in International Energy Policy at GSM London, University of Plymouth, UK. He is Professor of Construction and Contract Law at School of Engineering and Applied Science, Centennial College, Toronto, Canada, and a Principal Consultant at Envinergy Consult. He has published papers in international trade, environmental law and energy policy, and speaks at conferences on related issues. Kenneth Aidelojie completed his PhD in 2012 at the Centre for Environmental Policy, Imperial College London.

Adefolakemi Serifat Ayodele is a PhD research student in the Department of Sociology, University of Ibadan, Nigeria. She has a multidisciplinary background in social and management sciences with interest in industrial development, virtual office, diffusion, and technology transfer.

Feargal Brennan is Director of the University of Strathclyde, UK, having previously held senior positions at Cranfield University and University College London. Brennan is a leading authority on the integrity of offshore oil & gas structures and development of offshore renewables, including wind, wave, and tidal stream. He has written over 150 papers in peer-reviewed journals and conferences.

Frédéric Coulon holds a chair in Environmental Chemistry & Microbiology at Cranfield University, UK, and is recognised for his internationally leading contribution on water, soil, and wastewater treatment, resource recovery and environment, and public communication of environmental science and engineering.

Augustine O. Ifelebuegu is Principal Lecturer in Energy and Environment at Coventry University and a senior fellow of the UK Higher Education Academy. He has over 20 years' work experience that spans the energy industry and the academia. His research is focused on energy technology and environmental remediation with particular focus on renewable energy, process safety management, water and wastewater treatment processes, and oil spill clean-up and remediation.

Ahmed Kitunzi Mutunzi is Lecturer of Business Administration and Strategic Management in the School of Business, College of Business and Management Sciences at Makerere University, Uganda. Kitunzi holds a Bachelor of Arts in Economics & Education from Makerere University,

Uganda; an MBA from Maastricht School of Management, the Netherlands; and a PhD in Management from the University of the Witwatersrand, South Africa.

Athanasios Kolios (Dip Eng, MSc, PhD, MBA, PGCAP, CEng MIMechE, FHEA) is Reader in Risk Management and Reliability Engineering and Head of Offshore Energy Engineering Centre at Cranfield University, UK. During his academic career, he participated in several projects related to the development, analysis, and optimisation of energy systems through a balanced portfolio of research council funded, industry sponsored, and specialist consultancy projects of value exceeding £14.6 million as PI/co-PI/co-I.

Joy O. Ogaji is the pioneer Executive Secretary of the Association of Power Generation Companies (APGC) in Nigeria. Ogaji holds a PhD in Law from Warwick University, Coventry. She practised law both in the UK and in Nigeria and was a member of the Presidential Taskforce on Power, a body set up to drive the reform of the Nigerian power sector. Ogaji has contributed to several technical publications and is a passionate national and international speaker on public sector reforms

Stephen O. T. Ogaji (CEng, FHEA, FNSE) is a graduate in Mechanical Engineering with a PhD in Gas Turbine Engineering from Cranfield University, UK. He is a fellow of the higher education academy of UK and a fellow of the Nigeria society of engineers. He is the head of department for gas at the Niger delta power holding company, Nigeria, prior to which he was course director of the internationally recognised thermal power master's programme at Cranfield University and later the technical adviser to the Honourable Minister of Power in Nigeria. He was a member of the Nigerian Presidential Task Force on Power (PTFP). Ogaji has contributed to the publication of over 100 technical papers in peer-reviewed journal and conferences as well as two engineering textbooks.

Peter Ojo received his BEng degree in Water Resources and Environmental Engineering from Ahmadu Bello University, Nigeria, and his Master's degree in Environmental Management from Coventry University, UK. He is a part-time lecturer and doctoral researcher at Coventry University. His research interests include renewable energy, water and wastewater treatment, and process engineering.

Tade Oyewunmi LLD, is a senior researcher and lecturer at the Centre for Climate Change, Energy and Environmental Law (CCEEL), Law School, University of Eastern Finland (UEF), Finland.

George Prpich is Assistant Professor of Chemical Engineering at the University of Virginia, USA. He is recognised for his contributions to environmental engineering, risk management, and environmental policy and decision science.

Kabari Sam is Lecturer in Environmental Management and Pollution Control at the Nigeria Maritime University, Delta State, Nigeria. He is experienced in stakeholder engagement, environment, risk communication, resource recovery, sustainable and risk-based management of contaminated land. He is a consultant in sustainable environmental management approaches and standards.

Mobolaji Shemfe joined the Centre for Environmental Strategy as a research fellow in June 2016. He obtained his PhD from Cranfield University in 2016, having previously obtained an MSc in Petroleum Refining Systems Engineering in 2010 from the University of Surrey. His doctoral research focused on the techno-economic and environmental life-cycle assessment of the production of advanced biofuels via fast pyrolysis of biomass and bio-oil upgrading using conventional oil refinery technologies. His PhD work was an integral part of the SUPERGEN Bioenergy Challenge: Development of fast pyrolysis-based technologies for the production of advanced biofuel funded by The Engineering and Physical Sciences Research Council (EPSRC) and led by Sai Gu.

Tosin Somorin leads the development of a novel sanitary combustor at Cranfield University, UK. She has a multidisciplinary background in Applied Science and Engineering, with interest in technology development and solutions. Her research is focused on thematic areas of waste conversion, resource utilisation, and technology qualification, particularly in biomass and bio-energy.

Ayodeji Sowale is a research fellow in Mechanical Systems, recognised for the ability to improve competent designs and efficiency. He has BSc, MSc, and PhD degrees in Mechanical Engineering. He is working on system integration of the Nano Membrane Toilet (NMT) project funded by Bill and Melinda Gates Foundation at Cranfield University.

Shailendra Vyakarnam holds an MBA and a PhD from Cranfield University, UK. He is the Director of the Bettany Centre for Entrepreneurship at Cranfield and a resident senior member of Wolfson College, Cambridge. He is a fellow of the RSA (Royal Society for the encouragement of Arts, Manufactures, and Commerce) and the European Academy of Sciences and Arts. His particular focus is on science and technology-enabled entrepreneurship.

List of Figures

List of Tables

1

Introduction to Energy in Africa: Policy, Management, and Sustainability

Sola Adesola and Feargal Brennan

Energy is an integral part of a modern economy due to the important role it plays as an important factor of production employed in the production of goods and services, when combined with capital and labour (Keho 2016). It is also a significant contributor to global warming due to greenhouse gas emissions (Kolawole 2017). Africa's energy sector is vital to its future achievement of socio-economic growth and development. This is in sharp contrast with the abundant energy resources available, which could be harnessed to provide the needed energy. To achieve sustainable economic growth and development in Africa, affordable, sustainable, and

---

The authors acknowledge the work of Dr Aisha Kolawole [Kolawole, Aisha 2017." An Empirical Analysis of Energy Demand in Sub-Saharan Africa", PhD Thesis, Oxford Brookes University], whose work formed an important part in the background development of this introductory chapter.

S. Adesola (✉)
Oxford Brookes Business School, Oxford Brookes University, Oxford, UK
e-mail: sadesola@brookes.ac.uk

F. Brennan
University of Strathclyde, Glasgow, UK
e-mail: feargal.brennan@strath.ac.uk

© The Author(s) 2019
S. Adesola, F. Brennan (eds.), *Energy in Africa*,
https://doi.org/10.1007/978-3-319-91301-8_1

1

reliable energy must be provided. The stakes are enormous. Achieving the economic, social, and environmental promise of the region depends on the ability of the stakeholders to develop the energy infrastructure.

Africa Energy Outlook

The overall energy landscape in Africa is changing. Africa has experienced more rapid economic growth than in the past, with a corresponding increase in the demand for energy. Keeping pace with energy growth and needs is priority for policymakers, to enable economic growth and extend access to modern energy to all. These are not easy challenges, where supply lags demand and over 600 million people in Africa still do not have access to electricity, and approximately 730 million people rely on traditional biomass (IEA 2014). The percentage population with access to electricity in North Africa, West Africa, East Africa, Central Africa, and Southern Africa are respectively 98%, 47%, 23%, 25%, and 43%. Gross domestic product (GDP) per capita is generally three to five times higher in North Africa, where less than 2% of the population is without access to electricity (IEA: World Energy Outlook 2014). In contrast, half of the population in West Africa and three-quarters in East Africa and most of Southern Africa lack access to electricity (although only 15% in South Africa lack electricity access). North Africa on average consumes eight times more electricity per capita than the rest of the continent, excluding South Africa.

Trade and Investment

Since 2013, the Sub-Saharan economy has more than doubled to reach $2.7 trillion (IEA 2014). Recent economic growth is attributed to a variety of factors, including a period of relative stability and security, improved macro-economic management, strong domestic demand driven by a growing middle class, global drive for Africa's resources, population growth, and urbanisation. Nigeria and South Africa are the largest economies so far accounting for more than half of the SSA, with Angola, Ethiopia, Sudan, and Ghana being the next largest. On trade and investment in the region, the role of China's engagement is notable, particularly in oil, gas, and other natural resources which account for 80% of China's imports from Africa (Sun 2014) from 2005 to 2011. This is nearly double the level of the European Union and that of the USA over the same period. China's increasing stake in oil and gas in Africa is well known in Angola, Chad, and Uganda. However, the interest is not restricted to hydrocarbons; Chinese companies are among the largest investors in renewables across the continent including major hydropower projects, solar, wind, and biogas (IEA 2014).

Urbanisation

In defining the Africa energy outlook, rapid demographic trend in Africa is the growing rates of urbanisation, another result of economic growth, and the focus on industrialisation. For instance, the urbanisation rate for Africa was 61% and annual population growth was 1.07% (CIA World Factbook 2011). Population growth has been split relatively evenly between urban and rural areas, in contrast to the strong global trend to urbanisation. Only 37% of the Sub-Saharan population lives in urban areas—one of the lowest shares of any world region—which has important implications for the approach to solving the energy challenges. Growth in urbanisation and changing population framework lead to an increase in the demand for energy access and infrastructure.

Market Environment and Infrastructure

Businesses in Sub-Saharan Africa (SSA) most frequently cite inadequate electricity supply as a major constraint on their effective operation. It is a widespread problem that affects both rich and poor countries in energy resources. Insufficient and inferior power supply has a large impact on the productivity of African businesses (Escribano et al. 2010). The power problem is hinged on poor infrastructure; many countries face difficulties in financing the needed infrastructure costs. The Programme for Infrastructure Development in Africa (PIDA) identifies a need for $360 billion programme of infrastructure investment through to 2040, spread across energy, transport, information and communication technologies (ICT), and trans-boundary water resources (PIDA is led by the African

Union Commission [AUC], the New Partnership for Africa's Development [NEPAD], and the African Development Bank [AfDB]).

Governance, Affordability, and Sustainability

The role of government in the energy sector of Africa is substantial. Governance is defined by Kaufmann et al. (2008) as the traditions and institutions by which authority in a country is exercised and encompasses such factors as the process by which governments are selected, monitored, and replaced; the capacity of the government to effectively formulate and implement sound policies; and the respect of citizens and the state for the institutions that govern economic and social interactions among them. Significant to the realisation of economic development in Africa is the establishment of more effective and robust systems of governance. Many governments have failed on corruption, poor regulatory and legal frameworks, weak institutions, or poor transparency and accountability to shield actors from economic shock. These failings are significant factors as hindrances to investment in the continent, and thus pose a key issue for the energy sector if they are to attract foreign investors. Some countries have made progress than others in improving energy sector governance, but this is far from complete. For instance, five out of the nine countries producing 100,000 barrels per day (kb/d) of hydrocarbon liquids have new petroleum law under consideration—these include Nigeria, Ghana, Gabor, Congo, and South Africa. According to IEA (2014), transparency and accountability will continue to be important features of energy sector decision-making designed to command public acceptance and international respect.

Energy affordability is an important issue affecting end-user energy prices and expenditure. Prices vary significantly across SSA depending on whether they are regulated, controlled, or subsidised. Where subsidies apply, they serve to support energy access for the poor, but these are often exploited by fuel smugglers into nearby markets with higher prices. For electricity, tariffs in SSA, though vary by country and by type of customer, are in many instances among the highest anywhere in the world.

On average, SSA electricity tariffs are between $130/MWh and $140/ MWh, with those for services and industries being 5% and 8% higher than those charged to households. In comparison, electricity tariffs in Latin America, Easter Europe, and East Asia are around $80/MWh (Briceño-Garmendia and Shkataran 2011). The inability to set electricity tariffs at reasonable costs is a major obstacle to the long-term sustainability of many utilities in SSA.

Outline of the Book

This book *Energy in Africa: Policy, Management, and Sustainability* brings together conceptual, theoretical, and empirical contributions from a diverse set of experts and scholars to provide an analytical and interdisciplinary approach to energy in Africa from the lens of policy, management, challenges of sustainability, affordability, and cross-cutting issues. The book is compiled to better understand energy in Africa through rich assessments of the global energy environment and security in the region, along with policy, management perspectives, and challenges of environment and sustainability. To date, little or no book has explored energy in Africa from the lens of management perspectives. This well-researched analysis is enriched by African and international perspectives and anchored in detailed country examples. Bridging local perspectives on energy management and sustainability, this compendium transcends multi-stakeholder views and interdisciplinary approach of energy in Africa. The book covers a broad range of energy, for example, oil and gas, electricity, and renewables.

Understanding the policy, management, and sustainability of energy requires a broad knowledge of the energy business and the global economy. *Energy in Africa: Policy, Management, and Sustainability* equips readers with theoretical and applied knowledge of international energy production. Through the interdisciplinary nature of this book, readers will improve their understanding between state, business, and international agencies in the industry.

The **primary objective** of this book publication is to provide a more holistic, broader but deeper perspective of energy in the Africa developing economies than is commonly found in other books on the subject.

The book is an essential reading for advanced undergraduate and postgraduate students studying and researching energy management and policy, international business, energy sustainability as well as for practitioners who want to increase their understanding of such markets.

The book follows a **three-part structure with nine chapters** beginning with an introduction to the theme in Chap. 1, and then divides the chapters into parts and chapters, ending with a conclusion. The book covers a wide range of intellectual and practice-based scholarly thinking on energy policy, protectionism and regulation, stakeholder engagement, cooperatives' entrepreneurial potential to diffuse energy source, energy security, clean technology, and innovation, and issues on sustainability and affordability. The main chapters contain numerous citations of the main texts, which, together with the extensive reference section at the end of this book, should enable energy researchers and practitioners, as well as policymakers and students, to delve more deeply into each of the issues discussed.

Part I: Policy and Cross-Cutting Issues

Brexit and Trumpism: The Renaissance of Protectionism and Nationalism and Its Effect on the Twenty-First-Century Africa Energy Policy

Kenneth E. Afe Aidelojie

Chapter 2 explores the recent shifts towards the more protectionist and nationalistic policies in the political terrain of the two major energy-importing and consuming nations, the US and Britain, set to pose existential concerns and questions for energy and trade policymakers in oil-producing—especially those of Africa—developing economies. Understanding these shifts within an historic context, in relation to international trade laws in energy, and the domestic energy policy dynamics of Africa oil-producing states will be considered in this chapter. Also, appropriate recommendations on how policy blunders can be mitigated by policymakers of these rentier economies.

The Impact and Role of the Regulator on Electricity Market Growth and Sustainability in a Developing Economy

Joy O. Ogaji and Stephen O. T. Ogaji

Regulation is imperative, especially in the electricity sector, since electricity is a utility which serves the society with vested public interest. Chapter 3 adopts the exploratory methodology to review the role of the regulator in ensuring sustainability of the gains of privatisation in a developing country such as Nigeria. The chapter argues that the regulator can only be efficient and play its critical role, if freed from the political interferences from which it suffers currently. Several inhibitors of the regulatory process in the Nigerian power sector are highlighted, along with a review of the emerging functions of a well-versed sector regulator. The authors take the position that flexibility and incentive-driven methods should be used in regulating the sector, to avoid rigid regulation which may lead to stunted growth (Baumol 1996).

Part II: Management and Cross-Cutting Issues

Stakeholder Engagement and Sustainable Environmental Management of Oil-Contaminated Sites in Nigeria

George Prpich, Kabari Sam, and Frédéric Coulon

African nations are experiencing rapid economic growth and development, particularly within the energy sector; however, this growth has come at a cost to the environment and society. Nowhere have these impacts been felt more precisely than in the oil and gas producing regions of Nigeria where years of neglect and mismanagement have resulted in vast areas of hydrocarbon-contaminated lands. In Chap. 4, we present a case study of the Niger Delta. We show how constructive stakeholder engagement can be used to integrate the values and perspectives of affected communities and how this information can be used to inform environmental regulation and sustainable development. Lessons learned are relevant to other countries seeking to develop their energy resources in a sustainable manner.

Cooperatives' Potential to Diffuse Appropriate Solar Technologies in Uganda

Ahmed Kitunzi Mutunzi and Shailendra Vyakarnam

Chapter 5 explores the barriers and possible enablers for spreading solar technologies in Uganda. We focus on the requisites for solar diffusion and the relative strengths and weaknesses of Savings and Credit Cooperatives (SACCOs) in spreading solar technologies. The study is inspired by the apparent significance of renewable energy sources, the scarcity of electricity, and the documented attributes of SACCOs. The chapter is the result of a qualitative survey that entailed extensive literature reviews, focus group discussions, and interviews with various stakeholders of cooperatives. The chapter contributes towards viable policies for scaling up the uptake of solar-powered technologies and is framed in the context of innovation adoption theory.

The Evolving International Gas Market and Energy Security in Nigeria

Tade Oyewunmi

Over the past two decades, natural gas has increasingly become the preferred primary source of energy, especially for electricity generation. Developing countries with significant gas reserves, such as Nigeria, had previously dealt with gas mainly as a revenue earner through export projects. However, due to the growing demand for electricity and energy that trails social and economic development, there is now a significant move towards optimising domestic gas utilisation and supply. Thus, the regulatory framework for access, production, and commercialisation of gas resources, as well as safeguarding timely investments in supply infrastructure, inevitably becomes crucial to energy security. In view of such factors,

Chap. 6 examines the institutional and regulatory dimensions of energy security in Nigeria within the context of the evolving international gas supply market.

Part III: Energy Transition—Clean Technology, Sustainability, and Affordability

Clean Technologies and Innovation in Energy

Tosin Somorin, Ayodeji Sowale, Adefolakemi Serifat Ayodele, Mobolaji Shemfe, and Athanasios Kolios

Developing countries are faced with multiple energy challenges—the dilemma of increasing energy services to billions of people who currently live without electricity, and the need to operate low-carbon, intensive energy systems for environmental sustainability. Clean energy technologies can reduce fossil fuel dependency, provide jobs, and play a central role in improving access to energy; however, there are questions on availability, accessibility, reliability, affordability, and appropriateness of these technologies in developing countries. Chapter 7 provides an overview of the clean energy sources and technologies in Africa. It concludes with case studies of clean energy projects (potential, ongoing, tried and tested) across the different African countries and highlights the challenges, barriers, and approaches to the development, transfer, and diffusion of new and innovative energy technologies.

Renewable Energy in Africa: Policies, Sustainability, and Affordability

Augustine O. Ifelebuegu and Peter Ojo

Africa is endowed with significant renewable energy resources: abundant biomass, wind, hydropower, geothermal, and solar energy. However, these huge potentials remain largely unexploited, with Sub-Saharan Africa having the world's lowest electricity access rate, at only about 24%. The level of investment and policy interest in renewable energy remains low. A major challenge to Africa's efforts to develop renewable energy sources and the associated technologies is cost. Affordability is therefore a major consideration in the development/uptake of renewable energy technologies and depends mainly on the type of technology, policy direction, sustainability, and investment considerations. Chapter 8 will review the policies and policy direction of African countries on renewable energy and the factors that affect the affordability of the technologies currently and in the future.

Conclusions and Critical Perspectives on Africa's Energy: Can Multiple Perspectives on Energy Management Inform Policy and Practice?

Sola Adesola and Feargal Brennan

In this concluding chapter (Chap. 9), an attempt is made to identify common and emerging issues of energy across Africa countries as well as highlighting industry- and country-specific issues. Understanding the directions in which Africa's energy sector is set to develop is essential for policymakers and industry practitioners.

References

Baumol, W.J. 1996. Rules for Beneficial Privatisation: Practical Implications of Economic Analysis. *Islamic Economic Studies* 3 (2): 1–32.

Briceño-Garmendia, C., and M. Shkataran. 2011. *Power Tariffs: Caught Between Cost Recovery and Affordability*. Washington, DC: World Bank.

Escribano, A., J. Guasch, and J. Pena. (2010). *Assessing the Impact of Infrastructure Quality on Firm Productivity in Africa: Cross-Country Comparisons Based on Investment Climate Surveys from 1999 to 2005*. World Bank Working Paper Series: 5191, Washington, DC.

Kaufmann, D., A. Kraay, and M. Mastruzzi (2008). Governance Matters VII: Aggregate and Individual Governance Indicators for 1996–2007. World Bank policy Research Paper no. 4654.

Keho, Y. 2016. What Drives Energy Consumption in Developing Countries? The Experience of Selected African Countries. *Energy Policy* 91: 233–246.

Kolawole, A. 2017. An Empirical Analysis of Energy Demand in Sub-Saharan Africa. PhD Thesis, Oxford Brookes University.

Sun, Y. 2014. *Africa in China's Foreign Policy, Brookings.* www.brookings.edu/~/media/Research/Files/Papers/2014/04/Africa%20china%20policy%20sun/Africa%20in%20China %20web_CMG7.pdf. Accessed 30 July 2014.

The CIA World Factbook. 2011. Available at www.cia.gov/factbook.

Part I

Policy and Cross-Cutting Issues

2

Brexit and Trumpism: The Renaissance of Protectionism and Nationalism and Its Effect on Twenty-First-Century African Energy Policy

Kenneth E. Afe Aidelojie

Introduction

The US energy policy remains a major force in shaping the direction of global energy policies. A 2016 US Energy Information Agency report concluded that the country imported 3.2 million barrels of crude oil per day (bpd) from the Organization of Petroleum Exporting Countries (OPEC) States with 13% of this coming from African Member States. A rentier economy, that is, Nigeria, exported 210,000 bpd to the US in 2016. As Nigeria generates a significant amount of her budgetary revenues from hydrocarbons exports to the US and Britain, the need to strategically reconsider energy trade policies relative to these major trading

K. E. Afe Aidelojie (✉)
Department of Energy and Procurement, GSM London,
University of Plymouth, Plymouth, UK

School of Engineering and Applied Science, Centennial College,
Toronto, ON, Canada
e-mail: kaidelojie@live.co.uk

© The Author(s) 2019
S. Adesola, F. Brennan (eds.), *Energy in Africa*,
https://doi.org/10.1007/978-3-319-91301-8_2

partners, considering President Donald Trump's protectionist and nationalist grandstanding during his electoral campaigns and the results of the Brexit 2016 referendum, makes it now even more imperative for energy trade policy consideration.

Recent shifts in the direction of more protectionist and nationalist policies in the political terrain of these two major energy importing and consuming nations on both sides of the Atlantic (the US and Britain) are set to pose existential concerns and questions for energy and trade policy-makers in oil-producing, developing economies. Understanding these shifts within an historical context in relation to international trade laws of energy and the domestic energy policy dynamics of some of Africa's oil-producing States will be considered in this chapter. Historical protectionist energy and trade policy initiatives of the US will be reviewed; current relevant debates related to both countries' energy security, the consequences of pursuing the energy independence agenda, and potential trade and geopolitical considerations will be analysed. Also, appropriate policy recommendations on how policy flubs can be mitigated by policymakers of these rentier economies to minimise the effects of the US and UK protectionist drives fanned by the current sway of nationalism.

Fundamental to this chapter are, broadly, the following three narratives. The first is articulating the similarities of historic policy initiatives to the current renaissance of protectionism and nationalism energy trade agenda in the US and Britain, while using the analysis of these past initiatives to inform the importance of avoiding or harnessing the consequences of protectionism and how this will potentially inform the direction and shape the energy and trade policies of Africa's energy-rich countries. Second is what challenges are likely to confront oil-producing countries of the sub-region if the protectionist agenda is implemented and what ways can international trade laws on energy be effectively used to halt the sliding of nations back into the pre-trade liberalisation era. Finally, as the energy trade sector constitutes the largest primary commodity of global trade, most developing economies with large deposits of hydrocarbons sustain their budgetary revenues from the sale of the resource to the US and Britain; so what substantive policy initiatives should be considered by the developing economies to mitigate the possible effects of a protectionist energy policy from its major trading partners.

Organization of Petroleum Exporting Countries

The major oil-producing nations of Africa belong to OPEC. Central to this intergovernmental institution's overarching regulatory framework is Article 2 reinforced by Article 5 of the Charter of Economic Rights and Duties of States. Article 2 of OPEC's statute emphasised the fundamental pursuit of the coordination and unification of its Member States' petroleum (energy) policies[1] with the intended goal of influencing the international oil prices[2] to the benefit of its Member (oil producing) States.[3] This statutorily expressed intent fundamentally fuels the public debate for those arguing that OPEC and its Member States have a protectionist agenda and oversees a cartel to the detriment of international energy market liberalisation. This position has been referenced in different energy policy discussions by various Western governments that themselves have historically pursued and justified protectionist and nationalist policy tendencies on the grounds of national security, and there has evidently been a re-emergence in recent times.

OPEC's pursuit of a protectionist agenda—on the wings of a surging nationalist ideology—especially on the trade of its energy resources is definitely not new, neither can the West deny its involvement in such policies in the past. However, what seems to be significant and peculiar, post the thriving of market liberalisation and globalisation principles, is where the reversion to protectionist and nationalist ideologies have recently been coming from. Countries such as the US and Britain are historically considered frontrunners and strong proponents of globalisation and market liberalisation.

As oil remains the world's most valued commodity with the exploration sector contributing an estimated 4.6% to the global gross domestic product (GDP), the importance of oil to the global economy becomes even more obvious when a relatively small, fractional price change determines the direction of national and global economies. Although with a membership of 12 countries, OPEC countries supply about 40% of the world's oil, recent data suggest OPEC States and Russia hit a 2017 production high of 32 million bpd in July (Gloystein 2017). The immediate and long-term OPEC Member States' strategic direction remains critical to the global supply and price of energy, as recently shown by the OPEC decision to retain its production quota in the face of oversupply and slump in global oil prices. This continues to accentuate the importance of trade relations between the world's leading energy consumers and producers.

Historically, OPEC States have flexed their energy production muscles with obvious dominance in the area of energy resource reserve capacity and supply, due to their proven reserve and spare capacity, by restricting production and refusing to supply to those considered political foes. The socio-economic consequences on oil-importing nations, such as the US and Britain, whose economies are heavily dependent on supplies from these nations have been severe. According to Wagner (2002), the height of OPEC's relevance and importance was the Arab Oil embargo of 1973–1974, which resulted in the global energy crises of the mid-1970s. However, the resulting effect of the oil embargo was that affected countries, mostly in the West, developed policy initiatives that encouraged investment in research and development aimed at fast tracking the development and improving the efficiency of alternative sources of energy, developed once unprofitable marginal fields despite the cost implications, albeit mitigated by high oil prices, and averted any possible future hostage holding resulting from similar acts of embargo.

OPEC's decision in 2014 to aggregate its market share by pursuing a production strategy, which emboldens Member States to maintain or increase production in the face of oversupply, had a significant impact on the price of crude. According to the 2017 International Energy Agency (IEA) reports, the fall in oil prices upended the budget assumptions of all the producers, not just national companies entrusted with social and political obligations at home, but also their peers in the private sector. For OPEC countries, export revenues slumped to an estimated US $450 billion in 2016, down from US $1.2 trillion in 2012, causing major budgetary strains and in some cases making difficult political situations even worse. Also, with important impacts on the stock markets relative to the GDP of both exporting and importing countries, this could inform policymakers on the trajectory of future policies.

Africa States in OPEC

Africa has held a significant place in world oil supply over the last decade with a contribution to global crude production at over 12.1% and global gas production share at just over 10%, with significant reserve discoveries in Southern and Eastern African countries. In terms of consumption, the US Energy Information Administration's (EIA) short-term energy outlook for Africa also has an upward trajectory taking an almost 2.5% share of global energy consumption by 2018. Although many of the oil-producing countries in Africa are not members of OPEC, the contribution of African Member States to the organisation remains substantial.

Six African States are currently members of OPEC. According to the 2017 OPEC Annual Statistical Bulletin, the six Africa members (Algeria, Angola, Equatorial Guinea,[4] Gabon, Libya, and Nigeria) contributed 18.1% of the 25,014,000 bpd of crude produced, with a combined export value of $124,300 million. These nations remain rentier states whose budgetary revenue is dependent on revenues from their crude oil sale. A significant quantity of this crude production is exported to the US and Britain. An inward-looking energy trade policy initiative by these major crude oil resource consumers is bound to have a considerable socio-economic impact on developing OPEC Member States, especially those of Sub-Saharan Africa (SSA), with far reaching consequences on issues such as mass economic migration, which remains the main reason for the surge of a nationalist agenda.

Energy Trade and WTO: Bridging the Chasm of Protectionism[5]

The late twentieth century saw a concerted effort by nation states to do away with protectionist policies allowing for a more liberalised and inter-dependent trading regime using different economic and political instruments. An increase in multilateralism, globally championed by the US and Britain around the world, has resulted in the formation of transnational trade blocs and agreements such as the World Trade Organisation

(WTO), the North America Free Trade Agreement (NAFTA), and the Transatlantic Trade and Investment Partnership (TTIP). Increase in global trading has largely worked as a result of appropriate regulatory regimes and infrastructures put in place for the easy movement of many goods and services. Institutions such as the WTO have been at the forefront of pushing the non-protectionist agenda.

Energy goods and their related services continue to remain conspicuously outside the scope of the de-protectionist agenda of the WTO/the General Agreement on Trade and Tariffs (GATT). However, as argued by different authors, trade by its broader definition includes the cross-border sale of finished, intermediate, and capital goods, cross-border movement of persons essential for the production and sale of goods and services, the collection and spread of knowledge related to technology, process, and production methods, and so on, which makes energy resource a valid commodity to be included in the WTO/GATT regulatory regime.

Allowing oil and gas producing States of developing economies[6] to exercise sovereignty rights over their natural resources drawing on the fundamental principle of contemporary international laws[7] restricts the ease of energy resources' inclusion in the global trade regimes such as the WTO. Although there is an increase in the number of major oil and gas producing States acceding to the WTO, the strict interpretation of the sovereignty rights over these natural resources and the principle of non-protectionism under the WTO framework pose the main differential in principle and acceptance by both major exporting and importing nations. This difference has been reasoned by major oil and gas-importing States as inconsistent with the essence and spirit of the GATT, despite producing States holding their ground and rights to such anti-GATT protectionist practices through the provisions of Article 5, Charter of Economic Rights and Duties of States.[8]

With recent political developments in major countries known for advocating greater global trade liberalisation becoming more protectionist, uncertainties lie ahead. A wait-and-watch mode is definitely on in relation to how energy trade negotiations will pan out in future GATT/WTO Rounds.

Energy a Protectionist Commodity in International Trade

A recent IEA report indicates that 80% of global energy supply is composed of fossil fuel. Considering recent indications and a global resolve to engage in climate change mitigation strategies to deal with the impacts, energy trade is most certainly likely to be significantly affected in the foreseeable future. For instance, the Paris Climate Change Summit 2017, a follow-up to the 2015 Paris Climate Change accord, has seen commitments by major investors of significance (e.g. the World Bank) to withdraw their funding for fossil fuel exploration projects from 2019 which could have a significant effect on global supply (Elliott 2017).

An increase in the consumption of energy over the years for economic activities has significantly contributed to the overall advancement in living standards. An improvement in the liberalisation of energy trade has also been linked to increased economic activities, resulting in increased energy consumption (Sell et al. 2006). Trade in energy still has room for growth and improvement when considering the actualisation of its full trade liberalisation potential, especially within the plurilateral trading system, for example, the GATT/WTO.

Some major energy-producing countries are recent (or some are not yet) members of these trading systems (Selivanova 2011). Knowing that the GATT/WTO regimes as they are currently constituted do not adequately address the issues of energy trade, the push to integrate energy trade into these plurilateral trading systems is an evolving discussion (Marceau 2010), both in and outside the negotiation rounds, and requires the participation of energy-producing States. Broadening the membership and participation in negotiation rounds will be a step in the right direction to reduce resistance to the resulting framework and more importantly deal with the concerns that result in protectionist policies.

An energy (oil and/or gas) resource-rich country could also import the energy resources they are endowed with if the domestic demand outweighs the internal production capacity. For instance, the US is expected to have its crude oil production in excess of 10 million bpd by mid-2018, compared with the 2016 daily production of about 13.5 million b/d

(EIA 2017), despite being considered one of the world's oil-rich states. However, it is expected that through President Trump's "America First" political pledge and possible increased spending on infrastructure, energy consumption is expected to notably increase from the 2016 average daily consumption of 19.7 million b/d to over 28 million b/d in 2018. The high demand for oil to meet the domestic needs of the country will have to come from an increase in either internal production or importation from countries that have low demand needs but a surplus production— unless there is an increase in the domestic production for the domestic market. In this case, a country such as the US is considered to be an importing country.

Increased energy resources production resulting from advancements in exploration and production technologies and improved export capacity development strategies in the case of the US resulted in the so-called energy revolution. This has made the US become a net exporter of energy resource overall but also non-crude oil petroleum liquids and refined petroleum products—exporting small volumes of domestically produced crude oil, most of it to Canada, some of which may actually return to the US as refined products. While during the same period there was continuing decline in the import of oil from major oil-producing States such as Nigeria, according to an EIA 2012 report.

Producer nations that produce more than is locally required can export their energy resource to the international market. This will include most of the developing countries of SSA and other OPEC members. The overall local demand is less than the energy resource they have the capacity to produce; hence, the excess is exported to net energy importing nations.

As energy remains a critical factor in the advancement of human civilisation, realisation of national security is contingent on its sustainable, affordable, and reliable supply of this energy resource. To this end, the trade in energy resource as a commodity remains a substantial part of international trade. Broadly, its importance and relevance spread not only across the ease of transboundary movements but other economic, geopolitical and social issues, which various governments emphasise as necessities for national security. Holding a protectionist policy position could undermine the security interests governments seek to achieve. Different governments, not the least the US and Britain, base their foreign policy doctrines and domestic policy initiatives on their security of energy supply. Unfortunately, the impact of these on other economies, especially in terms of infrastructure destruction and overall impoverishing of society, is almost regarded as inconsequential when pursuing these foreign and domestic policy agenda.

Although the Obama administration is considered to have been more pro trade liberalisation and environmentally friendly (based on the multilateral trade treaties he signed the US into and the record number of environmental regulations signed into law), when compared with the current Trump administration, it was obvious that the dramatic increase in production due to exploration in shale oil favoured by technological advancements and the administration's energy policies continued to have an impact on the international energy market, even well beyond the administration. The oil glut consequent to oversupply led to the highest price decline of any commodity group (down 63% between June 2014 and December 2015). Although the oil price is rallying, it is too early to submit that recent world energy price challenges are behind us due to the policy initiatives of the Trump administration. There are no indications so far from the current administration that suggest that the interest of the oil and gas sector will be the opportunity cost for a more climate friendly environmental regime or an international trade model that is considerate of all and promotes market liberalisation. In fact, on the contrary, all indications show that exploration and investing in the sector will continue to increase in favour of domestic companies and international oil companies with host status in the US.

In terms of trade with Africa, the 2016 WTO report on trade showed that Africa's exports experienced a significant 30% decline in dollar terms in 2015 alone. This accounts for about 40% of the region's exports. Sub-Saharan oil-exporting countries, such as Equatorial Guinea and Congo, were significantly affected by a 60% decline in oil prices. Nigeria saw a decline of almost 50% in its export revenues in dollar terms. This weakness was also due to a variety of factors, including slow growth in North Africa resulting from domestic and political turmoil. Also, it was a response to the energy policies that encourage exploration and production of shale oil in major oil-importing countries (such as the US and Britain), not only reducing the export but also dropping the oil price due to a glut in the market. For instance, in West Africa alone, the dependence of the Economic Community of West African States (ECOWAS) on oil exports and imports—particularly Nigeria, accounted for 50% of ECOWAS exports—resulted in its share of world exports falling to 0.5% in 2015 from 0.9% in 2012.

Geopolitics and Protectionist Policy on Africa Energy Exports

Energy resource exporters in developing economies saw their exports drop primarily due to the increased global supply and falling global demand as a result of sluggish growth globally. The share of oil in developing countries' exports fell from 25% in 2012 to 21% in 2014. One reason for the decrease in fuel exports was the increased oil production by the US based on policy initiatives that allowed for the promotion of innovative use of technology in the exploration of once considered inaccessible oil and gas reserves. Between 2012 and 2014, the US reduced imports from Africa by 59% as a result of its increased domestic production. This decrease in energy imports contributed to a 47% fall in the value of Africa's total exports to North America during this period. In 2014, North America's share of Africa's total exports was only 7% compared with 11% in 2012. The budgetary ramifications of these sustained declines in exports, for many (if not all) of these SSA States, and along with other systemic inefficiencies continue to impact the socio-economic and infrastructural development of these countries. It can be argued that part of the underpinning reasons for the increase in domestic production in the face of excess supply is protectionist. Incentivising the domestic market with tax breaks and subsidies[9] continues to encourage production, even when it remains economically unviable. This sets to contribute to the decline in exports of the many African energy resource exporters.

As with other countries and regions, energy trade dynamics between Africa and the US/UK continue to be influenced by geopolitics. However, the strategic relevance of Africa is minimal compared to countries of the Middle East region. Energy trade decisions tend to be selective mostly in favour of oil-producing nations of the Middle East and North Africa (MENA) compared to the Sub-Saharan region. The involvement of the US and the UK in the MENA region is high and it is sometimes at the expense of peace and stability. Since the beginning of the twentieth century, energy is not only measured by its economic value but considering the resource and its derivatives within the context of important and strategic geographical sources and as a strategic political asset (UNCTAD 2000). The economies of Africa in this regard, like those of the Middle East, are of strategic importance fundamentally to ensure the success of the US/UK domestic economy and when or where it is perceived to threaten this fundamental interest, all measures are invoked to mitigate this. The use of targeted economic policies and military means to exert influence and achieve these are not uncommon. As mentioned above, the development and implementation of foreign policies by many countries are substantially aligned with these realities and the ability to play around the nuances in the world of diplomacy (and other means possible) gives the edge in trade deals.

Historic conflicts and bilateral or multilateral alliances can be linked to interests in energy resources, maintaining or increasing market shares, and creating alliances for ease of trade and transboundary movement. These events and alliances accentuate the importance and strategic significance geopolitics have on energy security and trade by underscoring national interest and security over all else. The increased push of a protectionist agenda in light of this could be viewed as a self-preserving strategy by the US and the UK to avoid the potential security of supply consequences by assuming an undue reliance on politically volatile and unstable regions where these resources are found.

Many of the SSA nations are notorious for being politically volatile with coups and counter-coups (although recently becoming more democratically stable), religious unrest and extremism, lack of infrastructure, kidnappings, terrorist havens, and a high propensity to hold and express anti-West sentiments not necessarily as a government policy in some cases but by citizens who may be willing to sabotage Western interests, resulting in socio-economic discontent, or from political or religious convictions. The implication is that all these potentially contribute to the justification for pushing protectionist measures pointing to the uncertainties that come with importing from these regions. Undoubtedly, these measures have an impact on the trade of energy resources between parties involved and more so, the economies that depend on its revenues.

It has been argued extensively by pro market liberalisation advocates that adopting restrictive energy trade policies pushes the issues of energy security to its brink and has a counterproductive impact on the global economy and the security of energy supplies. Social injustice and poverty promote hostility and social unrest. The singularity of the global energy market implies that a unilateral act or isolationist policy initiative by a State or group of States could have a considerable impact on the supply of or demand for available energy resources in the market, energy price, and, invariably, energy security (Leal-Arcas and Abu-Gosh 2014). Historically, events and policy shifts in major energy exporting and importing nations raise deep concerns for energy security. For instance, a policy push for energy independence is not cost-free. It potentially means an impact on the balance of payments and public debt that will have to be serviced by future generations of the country initiating the policy, as well as countries that will be denied the revenue from the exports of the energy resource to further develop their infrastructure.

The impending danger remains that when economic nationalism, as is currently re-emerging, is not adequately addressed, it slides over into outright protectionism in the hope that isolationist approaches and policies will provide viable answers to the reasons for the political surge of nationalism; however, this may be one wish too many.

US Historic Energy Protectionist Agenda

Energy policy in the US pre-dates to the colonial era, when the use of standing timber and coal for industrial and domestic use was encouraged and supported by the State. The discovery of oil and Edwin Drake's 1859 invention of deep oil well drilling technology exponentially furthered the oil age with the ease of commercial drilling. The emergence of oil as a major source of energy in the rebuilding post-World War II was driven by different factors, not least the energy intensity of oil, relative cleanness, and the significant improvement in exploratory and production technologies.[10]

There was growth in the exploratory capacity of the Consortium for Iran, commonly referred to as the "Seven Sisters". Advances in technology, championed by investment from international oil companies, such as Standard Oil, Texaco, and Gulf Oil, allowed for an easy and cheap supply of oil to their parent State from the oil-rich, mostly Arab, States; however, these companies had the support of their parent nations (with favourable fiscal regimes and economic policy initiatives), allowing for the provision of the energy needs of these countries. There were, however, concerns raised by different institutions (such as the Independent Petroleum Association of America [IPAA]) to the US government that increased reliance on the importation of oil from foreign States was a national security risk, not only economically but also militarily.

Over the years, oil import and national security have always been inextricably interwoven. Protectionist agendas have been pushed, especially in the areas of energy trade, by referring to and accentuating concerns around dependence on foreign supplies and the effects this could have on US national security. According to Bialos (1990), the Trade Expansion Act of 1962 Section 232 and its predecessor statute were used by various Presidents to address energy security concerns. President Eisenhower in 1959 on the premise of and concern for national security, and following the recommendation of the Director of the Office of Defence and Civilian Mobilisation (ODCM) said:

> "...crude oil and the principal crude oil derivatives and products being imported in such quantities...", "[...] threaten to impair the national security..."

Upon approval of Congress, the President was effectively allowed to use appropriate government instruments at his disposal to moderate imports of these commodities. The expansion of the interpretation of Section 232 allowed for the enactment and implementation of the Mandatory Oil Import Program (MOIP) in 1959. The consequent creation of OPEC[11] in 1960 and the collective action by some of its members in 1973 against the West following the Yom Kippur War inadvertently resulted in a major energy crisis and the MOIP was enacted to avoid a national energy crisis by increasing domestic refinery capacity while also stimulating domestic exploration capacity. This was done by allowing the imposition of import licences and quotas.[12]

The protectionist MOIP measure was designed and enacted to give preference to importation between US and Canada production. According to Economides and Oligney (2000), the implementation of the statute contributed significantly to 29% domestic production growth between 1959 and 1968. Considering this domestic production growth in parallel with the then cheap foreign oil mainly from the Gulf States, it could be safely assumed that the programme was a success. However, the increase in oil consumption substantially undermined the overall objective of the MOIP, resulting in subsequent Presidents needing to further push the protectionist agenda by increasing the domestic production quota.

With Section 232, President Nixon made the next major sweeping reform to MOIP in 1973 by introducing a graduated schedule of licence fee for crude oil and petroleum product importation, redefining the quota system—in favour of domestic producers—and suspending tariffs on petroleum products.[13] However, his decision, along with other leaders of Western countries (including Britain), to support Israel in the Yom Kippur War of 1973 effectively resulted in Arab Member States of OPEC sanctioning an embargo on major oil-importing developed nations. In the US, this resulted in a decline of about 2.4 million b/d of oil imports—an equivalent of 38% of domestic imports to the US (Bialos 1990).

Successive governments, pursuant to Section 232 provision and other regulatory provisions, pursued a series of policy initiatives also aimed at reducing US dependence on foreign oil. Although not outright protectionist, such initiatives had an extensive market interventionist undertone which encouraged an entitlement programme for small-scale oil refiners, and investment initiatives that impacted domestic oil production—ultimately controlling the price. As the US oil consumption consequently increased and dependence on foreign oil inevitably increased by 45% with a decline in domestic production (Energy Review, 1986), a supply crisis was almost inevitable. With the recent past experiences of the effects of the Arab oil embargo and the looming unrest in the Persian region (finally culminating in the 1979 Iranian crisis), the need to establish a Strategic Petroleum Reserve (SPR) in line with the recommendations of the IEA was inevitable but further dependence on OPEC rose significantly. However, during the early years of the President Reagan administration, the administration continued to deregulate the energy sector and promote the free market principles. Efforts were put in place to facilitate the revocation of President Eisenhower's oil importation licencing framework and revert the Carter oil control mechanisms[14] alongside other measures. President Reagan used the Executive Order 12287 to repeal the Price Control Program. In his statement, he emphasised:

> *"[...] price controls have held U.S. oil production below its potential, artificially boosted energy consumption ..." He further argued, "Price controls have also made us (the United States) more energy dependent on OPEC nations".*

This he insisted has negatively impacted on the nation's economic security and price stability.[15]

The argument was premised on the obvious reduction in the vulnerability to supply disruption resulting from previous governments having initiated market freeing and conservation incentives policies and the underpinning emphasis was the national security. This direction of nonrestrictive, energy trade restriction removal receded in 1982 due to concerns about the reliability of Libyan supply. This was also under serious scrutiny by the Reagan administration resulting in the invocation of the relevant legal provision (Section 232) for banning the importation of crude oil and refined oil products from Libya using Executive Order 12,538[16] (Bialos and Juster 1986).

By late 1979 oil production peaked while the oil price peaked in 1981 and steadily dropped until 1986. This latter year witnessed a dramatic drop in the market price from $23.29 in December 1985 to $9.85 in July 1986.[17] As the supply shortages disappeared, the consequence was a drop in energy prices. Overall, this could be attributed partly to the policy initiatives of the Reagan and subsequent Bush administrations in the US. It is important to note, prior to and during most of the Reagan-Bush administration, the core drivers of energy policies were "energy and economic security" with the overriding interest for national security. An increased awareness about air pollution resulting from the combustion of fossil fuel, and the potential impact on national security, increased the overlap between energy and environmental policy initiatives, which gained further prominence in successive administrations. This will not be discussed in detail in this chapter.

Successive governments have continued the agenda of making America energy independent, increasing internal production capacity by inadvertently engaging in policies that contribute to the set-out goal of pursuing the doctrine of energy security, national security, or investment in research and development, which ultimately has an impact on the export drive. During the Obama regime, despite the acclaimed environmental credential and policy initiatives, and significantly not signing to the Keystone XL Pipeline, the US experienced one of the highest production eras in its history. Over the last few years of his Presidency, between 2010 and 2015, US oil production grew at a rate that has few parallels in the history of the industry. In 2009, it averaged 5.4 million barrels of crude per day. In the month of March 2015, it was 9.4 million, approaching the all-time high of a little over 10 million reached in 1970 (Crook 2015).

Trump's Energy Dominance Agenda

There has been a further steady increase in the oil production capacity of the US since President Trump took office, according to the EIA December 2017 monthly report. In terms of oil drilling capacity, the seven US oil production regions collectively are expected to see an increase of 94,000 bpd in their drilling productivity capacity which could boost the nation's daily total production capacity to 11 million bpd in 2018 compared to 9.78 million bpd in 2017 and above the all-time high record of 10 million bpd in 1970. Although oil and gas are sold in an integrated global market, the US, compared to most oil-rich African countries, does not export as much crude oil; however, it exports many refined products, such as diesel fuel and other derivatives due to its extensive refinery capacity.

America's rising crude production continues to displace imports from Africa and the Middle East due to increased deregulation of the energy sector and revocation of orders from the Obama administration by the Trump administration and increased push to lease out Federal lands for exploration. With a negative forecast for production in many OPEC States, especially those of SSA (except Nigeria), the overall impact on the budgetary revenues and balance of payments for these countries is set to be far reaching.

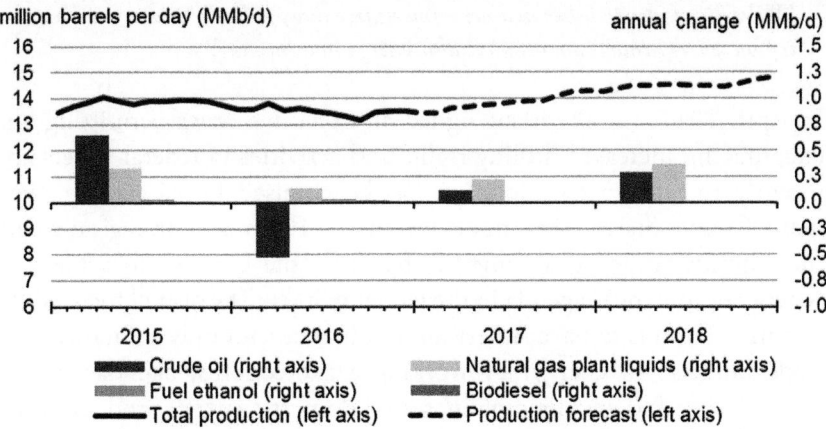

Fig. 2.1 US crude oil and liquid fuels production. (Source: US Energy Information Administration (EIA) Short-Term Energy Outlook Report, December 2017)

The May 2017 monthly oil market report of OPEC projected an increase in US shale oil production for 2017 to 14.45 million barrels per day, and US all-liquids oil production to increase to 15.2 million barrels per day in 2018 (see Fig. 2.1). This is expected to be further boosted with the increased supply from different potential offshore productions. However, it is important to contextualise this offshore supply. These developments are still very much in the early stages, with strong resistance from the Coastal States potentially able to halt the administration's ambitions.

On the surface, it seems that the Obama administration held stronger environmental credentials due to the burden of regulations placed on the hydrocarbon industry during his presidency, but in terms of the energy resource production, apart from coal, the other hydrocarbons also saw a substantial increase in production. Evidently, the current Trump administration is proving not to do anything different from his predecessors other than taking the American nationalist goal of "America First" a step further, from the goal of independence to that of dominance. In a recent speech in June 2017 during the Unleashing the American Energy event, the President articulated the administration's energy policy with the statement

"We're here today to usher in a new American energy policy", "...not only focusing on 'energy independence', but also 'energy dominance'".

In April 2017, the President signed an executive order permitting and encouraging increased drilling rights and activities in federal waters and recently the offshore exploration has been raised for discussion, with many Coastal States' Governors objecting.[18] This aligns strategically with the administration's goal aimed at fostering the US goal to achieve its "energy dominance" agenda in the global market. The overall focus of the administration is to have American dominance (not only militarily or in trade but also in energy) by working a protectionist agenda vis-à-vis incentivising and encouraging increased domestic energy production for internal consumption and net export.

Also, at other multilateral levels such as NAFTA, in the renegotiation under the mantra of "modernising NAFTA", a major consideration is clearly focused on seeking to integrate the North American energy market in such a way that will allow for America's dominance in the energy sector as it currently has in the service and manufacturing sectors.

The overriding interest of Trump's administration and the Republican-led government is centred on the creation of jobs for the American people. An increase in domestic production will help achieve this. In a statement from Paul Ryan, the Republican Leader of the House, published in the *Independent* newspaper, he praised the move to engage in drilling in Coastal States' waters as *"great news for the economy"* that would mean *"more jobs, more revenue, and more energy produced right here in America".*

UK Energy Trade Policy and Africa

In Britain, despite the surge in nationalist tendencies recently culminating in Brexit, the UK energy policy continues to evolve and reflect the changes in the international supply market, price variations, internal socio-political dynamics, environmental commitments, and different domestic supply issues. Successive governments have over the years laid emphasis on energy security; however, subsequent to the 23 June 2016 referendum to leave the European Union (EU), the refocus on energy security and the associated policies becomes even more vital due to the potential gap that will be created by leaving the EU. Although the UK in the EU had a relatively robust independent energy policy that continued to engage in committed energy trade with non-EU countries, especially suppliers from Africa and the Middle East even when it was a net oil and gas-exporting country, the resulting Brexit phenomenon will have a further significant impact on Britain's relationship with its non-EU energy trade partners. Regardless of Brexit negotiated outcomes, SSA energy suppliers are expected to remain key players in the future energy trade relations with the UK due to increased demand for energy for the British economy and its historic connections and trade relations with the countries in the region.

Considering the UK's continuing decline in its North Sea reserve capacity, the impact of Brexit on the country's energy costs will be more keenly felt according to a recent analysis by Stuart Elliott of S&P Global. Despite having been a net exporter fairly steadily between 1981 and 2003, Britain remains an import dependent, high energy-consuming nation since 2004, following a decline in the North Sea reserve capacity and production. The UK Department of Business and Industry Strategy data showed that for the third-quarter of 2016 imports made up some 39% of oil consumption, 62% of gas demand, and 65% of coal consumption in the UK (see Fig. 2.2). This is expected to remain as the demand for energy is forecast to increase based on the World Energy 2016 report. Traditional concerns related to oil and gas supply will persist; however, the degree to which this will escalate will depend on the outcome and the trade model Brexit negotiators reach with the EU and other non-EU energy trade partners. Currently, the energy trade between the UK and oil-producing Africa countries may be experiencing some degree of uncertainty due to the uncertainties associated with Brexit, as many Africa countries are having a cautious approach to trade negotiations with both the EU and the UK.

Although most of the trade arrangements the UK has with many nations were negotiated through the EU, Brexit has brought a new dynamic to trade arrangements and future relationships, especially with many of the SSA countries. The impact of this on energy trade is even

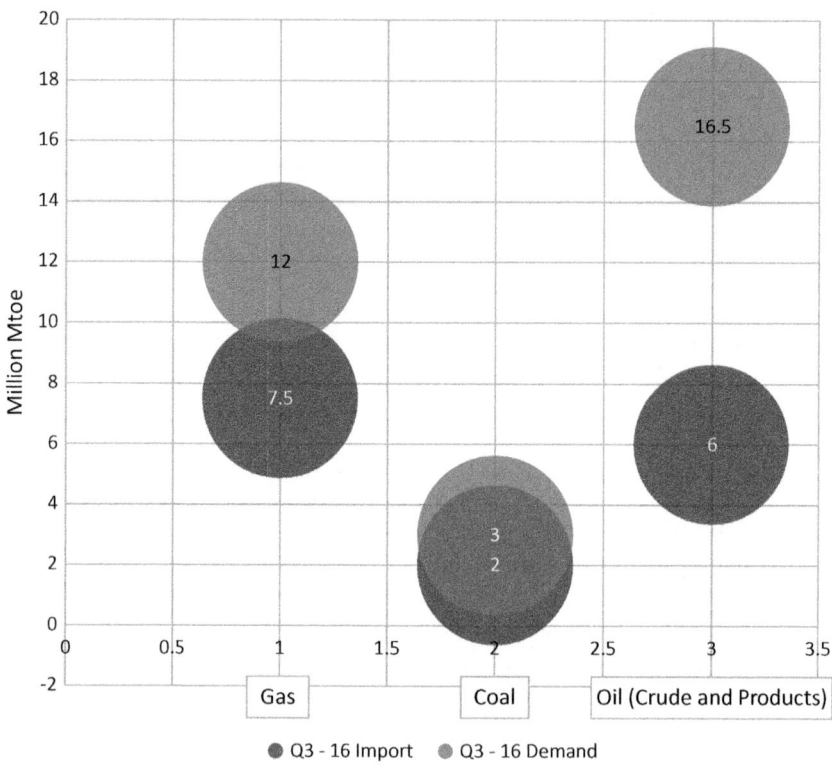

Fig. 2.2 UK dependent on energy imports in Q3. (Source: Adapted from UK Department of Business and Strategy 2016)

more uncertain due to the involvement of organisations such as OPEC (for Member States such as Nigeria and Angola) in the negotiation process. Extrapolating from previous stances of Britain on other trade issues, the silver lining remains. As pre-Brexit Britain was a strong voice for Africa in the EU trade policy discussions, especially with the Common Agricultural Policy, the process of redefining energy trade and diplomatic relationships could fall within the lines of a well-considered mutual economic benefit, knowing that this could potentially secure closer ties with other trading partners. A post-Brexit Britain may be willing to negotiate new bilateral deals with African nations that align the interests of both parties in a more beneficial manner.

Conclusion: Policy Recommendations for Oil-Producing African States

Overall, the interests of Africa will continue to remain peripheral for these countries; however, energy resource-producing countries need to harness the different leverages available to soften the effects of any possible protectionist resurgence from these energy trading partners. Africa has, and continues to remain on the receiving end of, nationalist and protectionist agendas of their developed nation trading partners. As is often the case, when there is a policy shift in many developed countries the impacts are significantly felt in developing countries, especially those in Africa. The need to anticipate these changes and proffer comprehensive responses is necessary to mitigate the negative impact on the socio-economic well-being of these countries. Nationalist and protectionist policies, though, seem to be becoming more obvious and prevalent in the US and Britain due to the influence of political populism. Historical accounts of US and UK energy policies validate protectionism as not new, especially using the grounds of national and energy securities as the rationale for pursuing this agenda. National security interests in many trade negotiations, domestic policy initiatives inter alia remain fundamental to every government. African States must start considering measures that will mitigate the impacts of protectionism on their major sources of revenue.

Developing and harmonising a robust set of local content laws among energy-producing nations of the continent will be of benefit to the nations. This should provide a collective base for a response to trade protectionist policies from trading partners and encourage the development of local capacity.

Increased intra-Africa trade has great potential that should be exploited to improve the socio-economic and infrastructural development of the continent. The need to revisit current energy policies that are focused on preferential engagement with non-African trade partners could provide the needed momentum for progress in developing the energy sector and broaden the market for the resource suppliers. Approaching this with the establishment of an Energy Free Trade Zone (EFTA) to incentivise investment in energy distribution network infrastructures, economies of scale, diversification, and value addition will result in the creation of domestic jobs on the African continent and improve the standard of living.

Africa must seek to lead in multilateral trade reforms that allow for greater trade reciprocity, especially in energy resource trade. The central premise of capacity building for the energy industry must be integral to the outcome of these negotiations.

The need to vigorously identify other markets with growing potentials is crucial for policymakers in Africa. The recent developments in Afro-Sino relations are positive; however, this is a short-term solution. The building of strong and credible institutions, according to President Obama, still remains the best and most sustainable way to reduce the reliance on the West or any other economic superpower. Africa must seek to develop itself in an economic power hub of relevance.

Notes

1. OPEC Article 2 (1): The principal aim of the Organization shall be the coordination and unification of the petroleum policies of Member Countries and the determination of the best means for safeguarding their interests, individually and collectively.
2. OPEC Article 2 (2): The Organization shall devise ways and means of ensuring the stabilization of prices in international oil markets with a view to eliminating harmful and unnecessary fluctuations.
3. OPEC Article 2 (3): Due regard shall be given at all times to the interests of the producing nations and to the necessity of securing a steady income to the producing countries; an efficient, economic and regular supply of petroleum to consuming nations; and a fair return on their capital to those investing in the petroleum industry.
4. Equatorial Guinea became a full member of OPEC on 25 May 2017.
5. The term protectionism has no universally accepted definition; however, there are some fundamental identifiers to that concept of protectionism. According to Altenberg (2016), few of these broad identifiers include: A set of policy instruments primarily affecting trade in goods; Policy instruments covered by the WTO or criteria based on WTO legality; Criteria based on whether public measures restrict trade; Criteria based

on whether public measures discriminate against foreign firms or other commercial interests; Criteria based on whether public measures distort markets; and Criteria based on the intent of the policymaker.

6. For the context of this chapter, the reference to oil and gas producing States will implicitly imply OPEC Member States, especially those of the Sub-Saharan region. Unless otherwise stated, Nigeria, Ghana (a non-OPEC State), and Angola will be at the core of this inference and information relevant to these nations will be used.

7. Permanent Sovereignty over Natural Resources, G.A. res.1803 (XVII), 17 U.N. GAOR Supp (No 17) at 15, U.N. Doc A/5217 (1962).

8. All States have the right to associate in organisations of primary commodity producers in order to develop their national economies, to achieve stable financing for their development and, in pursuance of their aims, to assist in the promotion of sustained growth of the world economy, in particular accelerating the development of developing countries. Correspondingly, all States have the duty to respect that right by refraining from applying economic and political measures that would limit it.

9. There is a varied estimation for US fossil fuel exploration and production annual subsidy. It ranges between US $10 billion and US $52 billion.

10. Progressively post-World War II, the US for instance maintained a healthy net export balance moving from 61,000 bpd in 1945 to 1,815,000 in 1960.

11. The resultant response by some major oil-producing States the following year was to the Organization of Petroleum Exporting Countries with the mission statement in Article 2 of their Statute.

12. Proclamation No. 3279, 24 Fed. Reg. 1,781 (1959).

13. Proclamation No. 4210, 38 Fed. Reg. 9,645 (1973).

14. Proclamation No. 3279, 24 Fed. Reg. 1,781 (1959)

15. The President, using the 1983 Proclamation No. 5141, 48 Fed. Reg. 56,929, retained all the oil tariffs established by the previous regulatory and policy initiatives.

16. Executive Order No. 12538, 50 Fed. Reg. 47,527 (1985).

17. This drop in price can be traced to Saudi Arabia's increased production in order to increase its market share, abandoning the OPEC strategic "cut production strategy" to increase prices.

18. The Governor of Florida, Rick Scott, recently secured a commitment from the Interior Secretary, Ryan Zinke, that Florida coastal waters will be exempt from the proposed leasing plan for coastal drilling for oil.

References

Altenberg, Per. 2016. *Protectionism in the 21st Century*. Stockholm: National Board of Trade, May – 1st ed. ISBN: 978-91-88201-13-3.

Bialos, Jeffrey. 1990. Oil Imports and National Security: The Legal and Policy Framework for Ensuring United States Access to Strategic Resources, University of Pennsylvania. *Journal of International Law* 11 (1990): 235.

Bialos, Jeffrey, and Kenneth Juster. 1986. The Libyan Sanctions: A Rational Response to State-Sponsored Terrorism? *Virginia Journal of International Law* 26: 799.

Bolton, Ronan, Antti Silvast, and Mark Winskel. 2016. Implication of Brexit for Energy Policy, Climate Change, Scotland's Centre of Expertise Connecting Climate Change Research and Policy (July).

Bordoff, J., and Houser, T, 2015. *Navigating the U.S. Oil Export*. New York: Columbia University: Center on Global Energy Policy. January.

Cottier, Thomas. et al. 2009. *Energy in WTO Law and Policy*. NCCR Trade Regulation Swiss National Centre of Competence in Research. Working Paper No. 2009/25.

Crooks, Ed, Financial Times. 2015. *The US Shale Revolution: How It Changed the World*. https://www.ft.com/content/2ded7416-e930-11e4-a71a-00144feab7de. Last Accessed on 28 Dec 2017.

Economides, Michael, and Ronald Oligney. 2000. *The Color of Oil: The History, the Money and the Politics of the World's Biggest Business*, Illustrated ed. (April), 168. Katey: Round Oak Pub.

EIA. 2017. *Forecasts A Mostly Balanced Oil Market 2018*. This Week in Petroleum, Release Date: December 13, 2017. Available at https://www.eia.gov/petroleum/weekly/archive/2017/171213/includes/analysis_print.php.   Last Accessed 10 Mar 2018.

EIA Report. 2012. *US imports of Nigeria Crude Oil Continued to Decline in 2012*. Available at https://www.eia.gov/todayinenergy/detail.php?id=5770. Last Accessed on 20 Dec 2017.

Elliott, Larry. 2017. World Bank to End Financial Support for Oil and Gas Industry. *The Guardian Newspaper*. https://www.theguardian.com/business/2017/dec/12/uk-banks-join-multinationals-pledge-come-clean-climate-change-risks-mark-carney. Last Accessed on the 17 Dec 2017.

Energy Review. 1986. *Energy Info*. Admin., 1986 Petroleum Supply Annual (1987); 1985 Petroleum Supply Annual, at 8 (1986).

Farah, Paolo D., and Elena Cima. 2013. Energy Trade and the WTO: Implications for Renewable Energy and the OPEC Cartel. *Journal of International Economic Law* 16 (3): 707–740 Oxford University Press.

Gloystein, Henning. 2017. *Oil Falls on High OPEC Supplies Despite Lower U.S. Crude Stocks.* Published Reuters Thursday, 3 Aug 2017. https://uk.reuters.com/article/global-oil/oil-dips-on-high-opec-supplies-defying-falling-u-s-crude-stocks-idUKL4N1KP185. Last Accessed on 20 Nov 2017.

Leal-Arcas, Raphael, and Ehab S. Abu-Gosh. 2014. Energy Trade as a Special Sector in the WTO: Unique Features, Unprecedented Challenges and Unresolved Issues. *Indian Journal of International Economic Law* 6: 1–53.

Leal-Arcas, Raphael, and Costatino Grasso. 2016. The Transatlantic Trade and Investment Partnership, Energy and Divestment. *Legal Studies Research Paper No. 213/2016 Published in Handbook of the International Political Economy of Energy, Palgrave, 2016.*

Marceau, Gabrielle Z. 2010. The WTO in the Emerging Energy Governance Debate. *Global Trade and Customs Journal* 5 (3): 83–93.

Selivanova, Yulia. 2011. *Regulation of Energy in International Trade Law: WTO, NAFTA and Energy Charter.* Alphen aan den Rijn: Kluwer.

Sell, Malena, et al. 2006. *Linking Trade, Climate Change and Energy,* 1–2. Geneva: International Centre for Trade and Sustainable Development.

Stuart, Elliot. et al. 2017. *Brexit – Commodity Implications: A Platts News and Analysis Feature.* Analysis: UK's Energy Costs Fuel Post- Inflation. Available at https://www.platts.com/news-feature/2017/oil/brexit-energy-implications/uk-energy-post-brexit. Last Accessed on 10 Mar 2018.

UNCTAD. 2000. *Trade Agreements, Petroleum and Energy Policies.* UNCTAD/ITCD/TSB/9. New York/Geneva: UNCTAD. Available at http://unctad.org/en/Docs/itcdtsb9_en.pdf. Last Accessed on 17 Nov 2017.

Wagner, Dewayne. 2002. The Decline of OPEC, Vol. 4. *Saskatchewan Economic Journal.* Published by University of Saskatchewan.

White, Jeremy B. 2017. *Trump Unveils Plan to Vastly Expand Offshore Oil Drilling in Latest Dismantling of Obama Legacy.* Published in the Independent. http://www.independent.co.uk/news/world/americas/us-politics/trump-offshore-drilling-expansion-plans-obama-climate-change-legacy-a8142741.html. Last Accessed on 5 Jan 2018.

WTO. *World Trade Statistical Review 2016.* Available at https://www.wto.org/english/res_e/statis_e/wts2016_e/wts2016_e.pdf. Last Accessed on 20 Dec 2017.

3

The Impact and Role of the Regulator on Electricity Market Growth and Sustainability in a Developing Economy: The Case of Nigeria

Joy O. Ogaji and Stephen O. T. Ogaji

Background

Several developing economies have all taken the path of reforming their power/electricity sector as far back as the early 1990s. It is also well known that these quests for reform by countries such as Nigeria were predicated on the associated disenchantment with the poor performance of state-owned power utilities, the desire for new investments and modernisation to meet rapid growth in demand, fiscal pressure, and the desire to protect and help the poor (Besant-Jones 2006).

The need for regulation in the electricity sector is predicated on two fundamental principles. Firstly, the fact that it is a public utility, and, as such, provides essential services for residential and industry, that is, the common good ("affected with the public interest"). Secondly, the setting

J. O. Ogaji (✉)
Association of Power Generation Companies, Abuja, Nigeria

S. O. T. Ogaji
Niger Delta Power Holding Company, Abuja, Nigeria

© The Author(s) 2019
S. Adesola, F. Brennan (eds.), *Energy in Africa*,
https://doi.org/10.1007/978-3-319-91301-8_3

provides an avenue for technological and economic features and enables monopolistic tendencies which may be inimical to market competitiveness. These two reasons make economic regulation imperative in ensuring public benefits/interest that the market may be unable to achieve on its own.

The overall aim of regulation is to ensure the provision of safe, adequate, and reliable services at prices (or revenues) that are sufficient, but no more than sufficient, to compensate the regulated firm for the costs (including returns on investment) that it incurs to fulfil its obligation to serve (Section 32 EPSR Act 2005). The essence of regulation, from as far back as the medieval period, is hinged on protecting public interests. The US Supreme Court had this to say:

> *Property does become clothed with a public interest when used in a manner to make it of public consequence, and affect the community at large. When, therefore, one devotes his property to a use in which the public has an interest, he, in effect, grants to the public an interest in that use and must submit to be controlled by the public for the common good.* (U.S. Supreme Court, *Munn v. Illinois*, 94 USC 113, 126 (1877))

The electric power sector's economic immensity and vastness in geographic scope, combined with its wide ownership and management, makes regulation imperative. The foregoing portend that Regulators are faced with the task of balancing several competing needs and objectives whilst focusing on optimising for least-cost service delivery and ensuring high quality and reliability (Bazilian et al. 2013). Zinaman et al. (2014) in corroborating this assertion highlight the key challenges of modern-day power sector regulation which involve reconciling the new regulatory objectives with the already difficult task of balancing existing objectives. The facts adduced above, in addition to the need for a more efficient role of the Regulator, necessitated this chapter in this book.

This chapter focuses on exploring the role of the Regulator in sustaining the gains of reform in a developing economy, with a focus on Nigeria. To do that, the chapter begins with a brief overview of the power sector, encapsulating the strategies that guided the process. The current challenges of the power sector and the expected interventions of the Regulator are evaluated. The chapter looks into market structure and restructuring power systems, drawing on experiences from other jurisdictions, and outlining the emerging role of the Regulator in the Nigerian power sector.

The chapter outlines some of the key features and challenges of the regulatory landscape of the Nigerian Electricity Supply Industry (NESI) and emphasises the powers and functions of the Regulator. The imperatives of the key principles of an effective Regulator, hinged on independence and transparency, as well as striking a balance between encouragement of investors and consumer protection, are also discussed. The chapter concludes by highlighting some critical facts in ensuring effective regulation as one of the panaceas in sustaining the gains of privatisation and by extension the electricity market growth and its sustainability.

Evolution of the Nigerian Electricity Market

Research shows that public electricity business in Nigeria began in 1896 with the firing of a 20 MW diesel plant in Ijora, Lagos. However, research also shows the existence of pockets of small diesel power plants commissioned to supply electricity to isolated cities in Nigeria. In 1951, the Electricity Corporation of Nigeria (ECN) was formed to co-ordinate and oversee the electricity distribution business all over the country using captive or embedded generators. In 1956, a 30 MW coal plant was fired at Oji River to add to the number and capacity of generating plants in the country (Awosope 2014).

Following increasing demand for electric power, the Niger Dams Authority (NDA) was set up in 1961, to explore, build, and manage dams for hydroelectric generation, in addition to building, operating, and maintaining the transmission grid in the country, and to manage other sources of power for the country, including gas, coal, oil, and so on. This led to the transfer of ownership of all existing generation assets to the NDA. The above arrangement, led to the establishment of two separate entities—the NDA was responsible for energy generation and transmission, while the ECN was responsible for energy distribution and marketing.

On 1 April 1972, ECN and NDA were merged to form a unified body known as the National Electric Power Authority (NEPA) with the actual merging taking place on 6 January 1973 when the first manager was appointed. A critical question to ask at this juncture is: What is the rationale for the bundling of the ECN and NDA?

The following are some of the reasons adduced as to why such a decision was made:

- To maximise the benefits of standardisation, for example, cost reduction
- To eliminate the duplication of managerial capacity and improved coordination of power supply
- To reduce operating costs and waste
- To eliminate bickering between NDA and ECN on payments for energy
- To optimally utilise the revenue from the sale of electricity to enhance the uniform development of the entire power system
- To uniformly remunerate workers from NDA and ECN after job evaluation, to enhance morale and productivity
- To optimise decisions on locations of generating facilities and their corresponding transmission and distribution (T&D) infrastructure around the country.

It is imperative to further note that the bundling led to massive network growth between 1975 and 1990, with four additional power plants built to add 3200 MW capacity to the National Grid. Total generating capacity of the industry by 1990 was about 6200 MW, while total route length of transmission lines was about 10,000 km.

Between 1990 and 1999 there was no significant investment in the industry—plants and equipment could not be adequately maintained, and system expansion and reinforcement could not be carried out. Decay set in, resulting in frequent power cuts, high technical and non-technical losses, high level customer dissatisfaction, and low revenue collection. The Federal Government had responsibilities in the other areas of the economy and could not provide all the investment and management needed to meet the growing demand for electricity. As a result, the Federal

Government sponsored two panels of enquiry to develop models for restructuring NEPA into an independent unit or towards privatising it out of its monolithic nature. This led to the establishment of the electrification boards whose work was to take power supply to the rural areas and new cities.

The monopoly status of NEPA was amended in 1998 to make way for the entry of the other categories of electricity service providers, including private enterprises, into the electricity market. In spite of this amendment, the private sector did not feel sufficiently encouraged to participate actively in the development of the industry. In 2001 the Federal Government of Nigeria (FGN) approved a power policy which proposed a reform of the power sector, to promote a sustainable electricity industry growth (NERC website).

The FGN, in implementing a viable reform process, had the following strategy in place to guide its work:

- Restructuring (Unbundling):

  - Separation of the monopolistic activities of the electricity industry from the potentially competitive functions
  - Creation of smaller, ring-fenced business entities that are easier for private sector participation

- Sequencing of the Unbundling in the NESI:

  - Functional separation—Separation of employees involved in different activities with separate managements (Jan. to Dec. 2004)
  - Accounting separation—Keeping separate accounts for different activities within the industry
  - Operational separation—Unbundling of investment decision-making process. Unbundled entities make their own investment decisions
  - Corporate unbundling—Creating separate legal entities whilst preserving common ownership. Registration of the successor companies with the Corporate Affairs Commission (CAC).
  - Divestiture or ownership separation—Separating different activities into distinct legal entities with different ownerships (Nov. 2013).

It is important to point out that the completion of the above unbundling did not signify the end of Reform, as Reform was considered a very long journey. Electricity Market Reform it should be noted is a process or a set of processes, certainly not an event.

The transition within the Nigerian Electricity Industry started in earnest in 2001 when the National Council on Privatisation (NCP) issued the Nigeria Electric Power Policy (NEPP). The policy argued that the collapse of the electricity sector in Nigeria would only be cured with the liberalisation of the sector. This was to create a competitive and efficient electricity market characterised by the existence of an independent Regulator, and utilities that are committed to cost efficiency and cost recovery.

The NEPP was premised on ensuring sustainable improvement in electricity supply through enhanced commerciality. The objective was to create a new electricity industry based on rules that are enforced by an independent Regulator. The Regulator was to be mandated to ensure that 'efficient operators recover prudent costs'. This cost recovery was hinged on efficiency.

The NEPP was encoded in legislation through the Electric Power Sector Reform Act 2005 (EPSR Act 2005), to enable the power sector reform to achieve institutionalisation. The Act formed the fulcrum of the reform, giving birth to 18 companies as follows: 6 legacy generating companies or GenCos, 1 transmission company, that is, the Transmission Company of Nigeria (TCN) or TransCo, and 11 distribution companies or DisCos.

The generating companies or GenCos are following:

- Sapele Power Plc Now G-Eurafric
- Ughelli Power Plc now Transcorp Power Ltd.
- Afam Power Plc yet to be privatised
- Geregu Power Plc owned by Forte Oil
- Shiroro now North South Power (NSP) as concessionaire
- Jebba and Kainji. Now Mainstream Energy Solutions Limited (MESL) as concessionaire.

It should be noted that Nigeria's largest power plant, Egbin Electricity Generating Company (EEGC), had previously been sold to a consortium of KepCo and Sahara, and hence was not part of the bidding process.

The 11 distribution companies or DisCos are following:

1. Abuja Electricity Distribution Company (AEDC)
2. Benin Electricity Distribution Company (BEDC)
3. Eko Electricity Distribution Company (EkEDC)
4. Enugu Electricity Distribution Company (EnEDC)
5. Ibadan Electricity Distribution Company (IbEDC)
6. Ikeja Electricity Distribution Company (IkEDC)
7. Jos Electricity Distribution Company (JEDC)
8. Kaduna Electricity Distribution Company (KdEDC)
9. Kano Electricity Distribution Company (KnEDC)
10. Port-Harcourt Electricity Distribution Company (PHEDC)
11. Yola Electricity Distribution Company (YEDC).

The Federal Government retains 100% ownership of the transmission company and 0% in all the generating plants, except in Geregu GenCo where it retains 49%. In the case of the distribution companies, the Government retains a 40% stake in all 11 privatised companies.

The NERC and Its Anticipated Roles in the Nigeria Power Market

Arguably, the role of the Regulator in market sustainability cannot be over-emphasised. In recognition of this fact, the Electric Power Reform Sector Act, 2005 gave birth to the Nigerian Electricity Regulatory Commission (NERC) on 31 October 2005. Section 96 (1) & (2) (c) & (d) of the Electric Power Sector Reform Act 2005 (Act No. 6 of 2005), empowers the Commission with unlimited powers to do all and what it deems fit to ensure efficiency in the electricity market. In other words, it means that NERC has been granted full regulatory authority over the NESI. The Commission consists of the Chairman's office and six Divisions: Market Competition and Rates; Legal, Licensing, and Enforcement; Finance and Management Services; Government and

Consumer Affairs; Research and Development; and Engineering, Standards, and Safety. NERC was mandated to oversee the orderly reform of the electricity industry and ensure adequate, safe, reliable, and affordable electricity to all consumers.

Furthermore, NERC was expected to perform the functions of:

- Determining the revenue requirement
- Allocating costs (revenue burdens) among customer classes
- Designing price structures and price levels that will collect the allowed revenues, while providing appropriate price signals to customers
- Setting service quality standards and consumer protection requirements
- Overseeing the financial responsibilities of the utility, including reviewing and approving utility capital investments and long-term planning
- Serving as the arbiter of disputes between consumers and the utility.

A key question the reader may ask is: Is NERC, the Nigerian power sector Regulator, able to perform and carry out its mandate as encapsulated in the NEPP 2001 and the EPSR Act 2005? The answer to this question we believe is negative. This is because the enabling environment for any policy to succeed in Nigeria and some developing countries, notwithstanding its success rate in any jurisdiction it is transported from, is lacking. Without this enabling environment, implementation of such imported practices, even in amended forms, becomes a Herculean task.

The reasons adduced in the following section will throw more light on why NERC, the power sector Regulator in Nigeria, is not able to effectively monitor and carry out its mandate and functions.

Inhibitors to Regulatory Certainty in the Nigerian Power Market

A key aspect of the power sector reform is to ensure the perpetration and perpetuation of regulatory certainty in the power sector. Without doubt, regulatory certainty is germane to the attainment of the objectives of the power sector reforms. To ensure regulatory certainty, the EPSR Act (2005) made provisions for the following:

- An independent regulatory Commission with clearly defined regulatory functions and objectives backed by law
- Clear process of appointment and removal of Commissioners and Chairman of the Commission outside of the normal Civil Service appointments
- Isolation of the Commission from Civil Service rules, including its reporting structure, compensation structure, and retirement benefits of its Commissioners and staff
- Autonomous funding for the Commission with direct appropriations from the National Assembly, outside of appropriations to the Ministry of Power.

In summary, the regulatory environment under the EPSR Act is designed in such a manner as to reduce the influence and interference of the Federal Government in the regulation of the power sector. The Act also takes away regulatory powers of the Minister and President in the power sector (Amadi 2014). Despite the good intentions of the EPSR Act in instituting independent regulation of the power sector, the sector is still faced with regulatory risks arising from regulatory uncertainties, Government's continued influence in the affairs of the Regulator and ineffective regulation of the sector by NERC. In a privatised market, Government's role should be focused on policy formulation, for example, opening of the domestic electricity market to foreign investment; however, there have been several occurrences that have escalated the regulatory risk in Nigeria.

A very good example is the incidence where the first appointed Chairman of NERC, Dr Ransome Owan and his fellow Commissioners were casualties of Government's interference in the power sector as they were removed by the Yar Adua administration without following due process (as set out by the EPSR Act 2005), well before the expiration of their five-year tenure. That misadventure by Government, coupled with the appointment of an Administrator for NERC, thwarted effective regulations in the power sector and stalled the power sector reforms for over two years, including the implementation of the Multi-Year Tariff Order (MYTO) tariffs, until the Federal Government, in tacit admission of its errors, reached an out-of-court settlement with Dr Owan and the other Commissioners. The regime of the next Chairman, Dr Sam Amadi, was also not without Government's undue influence over NERC's ability to regulate the power sector effectively, particularly in enforcing market rules and tariff setting (Odion 2016).

Regulatory risks can also be attributed to the behaviour of Market Operators and Licensees. A situation whereby Market Operators, Licensees and interested stakeholders seek to circumvent regulatory orders or do not agree with regulatory orders issued by the Regulator, but approach the Presidency or the Minister of Power to seek redress or to give backing to circumvent such orders, increases regulatory risks to the sector (Odion 2016).

Research has shown that regulatory independence is key to private sector confidence in the power sector and has a significant effect on the ability of Government to attract and sustain investments in the power sector. Prof. Guy Holburn of the Richard Ivey School of Business as quoted in Amadi (2014) notes that, "strong regulatory governance regimes consist of expert agencies that operate largely independently of day-to-day political control, but under legislative mandates and procedural requirements that safeguard the rights of stakeholders. Such regulatory regimes can provide credible assurances to industry and stakeholders that policies will not change in an arbitrary or unpredictable fashion, for instance in response to new political or economic pressures, after investments have been made."

Regulatory governance regimes that enhance agency independence from political intervention in day-to-day decision-making have the benefit of encouraging greater levels of private investment and at lower cost, which benefits consumers, since perceived regulatory risks are reduced. Amadi (2014) further notes that,

> weak regulatory governance is characterised by a more politicised policy-making environment, where the Government has greater control over regulatory policies. In this type of environment, it is more difficult to achieve credible commitment to future investor and stakeholder protection,

heightening perceptions of regulatory risk. In the absence of adequate regulatory governance, a jurisdiction may encounter multiple types of inefficiencies in its electricity sector – under-investment, minimisation of maintenance expenditures and fluctuating rates of capital investment. Ultimately, the negative effects of non-credible regulatory governance can lead to Government ownership becoming the default mode of operation.

Without doubt, more needs to be done in terms of effective regulation of the power sector in Nigeria. While there is a need for the Regulator to interface with Government in its function and duties within the limits of the EPSR Act and policy direction of the Government, it should be made clear that the Regulator must always be seen to have independently and transparently arrived at regulatory actions and orders. *"As we approach a critical phase of stabilising the power sector, the ability of the Regulator to regulate the industry without creating regulatory risks, following "orders from above" and caving in to the demands of Market Operators that are not fair to the consumer, is vital for the sustainability of the power sector reforms"* (Odion 2016).

The imperative role of regulation in any electricity/power industry is hinged on the fact that its lack will defeat the aim of any Government in liberalising that industry, as private behaviour, unregulated, will diverge from the public interest which is the norm for public utilities (Amadi 2014). It is rather unfortunate that the Nigerian power sector is bedevilled with politics, lack of relevant capacity, high handedness, and micromanagement of the "privatised entities," which have inhibited a complete and market-driven reform relative to market structure, degree of private participation, and development of the regulatory framework.

The Critical and Emerging Functions of the Regulator in Market Sustainability

Consistent with the dynamics and technological advancements in the value chain and the prospects of renewable energy (RE), such as balancing the challenges of demand and supply, appropriate generation mix coupled with the persistent pressure for ensuring sustainable economic development, increased access to regular and affordable electricity by consumers in the urban and rural areas, have all added up to escalate the already complex task of power sector regulation. Research has shown the balancing act that regulators in the power sector are faced with, to enhance value at least-cost service delivery, while ensuring high quality and reliability (Bazilian et al. 2013).

The importance of the Regulator in the electricity sector is:

• To constrain the exercise of monopoly power by incumbent suppliers
• To provide incentives for operating efficiency and quality of service
• To optimise the structure of the sector
• To promote least-cost system expansion (with private capital invested in independent power producers—IPPs)
• To stimulate energy conservation and research and development.

Although no component of the regulatory task in the power sector can be undertaken lightly, the most critical role of a Regulator in market sustainability is the ranking and synchronisation of various objectives that typify the vital challenge of power sector regulation. Arguably the emerging regulatory roles can be better comprehended through the framework of existing objectives and an additional layer of emerging objectives.

It is in corroboration of this fact that Zinaman et al. (2014) outlined the eight existing objectives of power sector regulation which are as follows:

• Design and Manage Electricity Tariffs
• Maintain System Reliability, Meet Demand Growth, and Expand Electricity Access
• Ensure Financial Health of Utilities
• Facilitate Private Investment
• Protect the Interests of the Poor
• Support Technical Safety and Reliability of the Power System
• Enhance Energy Security and Manage Risk

In what follows, we provide further context to some of these primary objectives of regulation.

Design and Manage Electricity Tariffs

One critical role of a Regulator in the electricity or power sector is to design fair and equitable electricity tariffs for the various customer classes, with a timeline for a review of those tariffs considering several building blocks, and hence periodically redesigning the tariffs considering a variety of objectives. In Nigeria, as in any other country, this is contained in Section 32 (d) of the Electric Power Sector Reform (EPSR) Act, 2005 with a major objective of ensuring that the prices charged by Licensees are fair to customers and sufficient to allow the Licensees to finance their activities and obtain reasonable profit for efficient operations (Section 76 (2) EPSR Act, 2005).

The above function is made possible in accordance with the authority given under Section 76 of the EPSR Act 2005, and to carry this out efficiently, the Commission established a methodology for determining the electricity tariff in the NESI and subsequently issued a Tariff Order called the MYTO that sets out tariffs for the generation, and T&D of electricity in Nigeria (NERC website).

The MYTO is a tariff model for incentive-based regulation that seeks to reward performance above certain benchmarks, reduces technical and non-technical/commercial losses, and leads to cost recovery and improved performance standards from all industry operators in the NESI. The irony of the reality is that in spite of the MYTO being an incentive-based regulation, meant to lead to improved performance in the industry, due to so many factors some of which have been discussed and others are yet to be discussed, the Regulator is incapable of handling critical issues of the sector, such as monitoring the operators to the industry standards. There has been endemic underperformance of the distribution companies who claim their inefficiency and lack of investment on infrastructure on the lack of a cost reflective tariff. The Regulator, due to lack of its capacity and influence from Government, has left the market with a heavy debt burden.

Maintain System Reliability, Meet Demand Growth, and Expand Electricity Access

This is yet another critical role of the Regulator. Electricity disruptions and supply shortages can have significant economic costs to a jurisdiction as a whole (Balducci et al. 2002; Sanghvi 1982). A significant thread that would most certainly unravel, isolate and lay bare Nigeria's problems, jump start unparalleled economic growth and ultimately lead to overall development, peace and simultaneous progress in all geopolitical zones of Nigeria, is constant and sustained electricity availability.

Achieving a constant and sustainable electricity supply is one of the key elements required to actualise the peace and prosperity of the citizens of any nation. This is a fundamental requirement for advancement, security, and progress of any country, including Nigeria.

The Regulator is tasked with assigning costs across rate classes to pay for such projects, as well as coordinating with Government and international development efforts to optimally execute electrification efforts. Section 96 (1) EPSR Act 2005 buttressed further the role of the Regulator. The continued, abysmally low supply of electricity accounts for the significant throttling of the economy. The average Nigerian's desire and willingness to pay an appropriate price for electricity consumed can be clearly seen in the great demand for generators of all shapes and sizes, and how much they pay to operate and service these generators. The imperative role of electricity to any economy cannot be over-emphasised; hence, it is asserted that about 70% of the challenges experienced by the country are caused by the Regulator's action or inaction.

Ensure Financial Health of Utilities

"Show me the money" is a well-known phrase used by Rod Tidwell (Cuba Gooding Jr.), while negotiating with his agent, Jerry Maguire (Tom Cruise), in the 1996 film, *Jerry Maguire*. This catchphrase typifies the current situation of the NESI. Currently, the flow of money within the industry is the fundamental problem preventing Nigerians from enjoying continued and sustainable improvement in electricity supply, and thus the gains of the Nigerian power sector reforms. Addressing issues regarding the financial viability of the electricity supply value chain is the main and immediate issue preventing Nigerians from enjoying the benefits of the Power Sector Reforms (Madu 2016).

This role also is enshrined in the EPSR Act as a critical role of the Regulator who, due to several factors enunciated above, is unable to carry out its functions. Consequently, improving the financial viability of the electricity supply value chain will ensure the improvement of the electricity service delivery and thus provide the required platform and goodwill to continue driving the industry reforms, since electricity customers are keen to see immediate improvements in service levels and will not be satisfied with the promise that improvements will materialise after the sector has been fully reformed.

In the short term, the current poor status of the electricity market has reduced and almost eliminated the ability of fuel suppliers, Generation companies (GenCos) and Transmission Company (TransCo), to meet their financial obligations to their vendors and suppliers, thereby threatening to completely undermine the entire electricity value chain. Obviously, it is not sustainable to keep power generating plants and transmission lines operational with the current level of payments that the value chain is currently receiving (Madu 2016). This is because GenCos and the TransCo of Nigeria have to cater for their liabilities, which include gas supply and transportation, long-term maintenance, procurement of expensive spares, remuneration of staff, servicing of bank loans, and payment of insurance premiums required to sustain their operations.

Strategically, given the fact that in the reformed NESI, projects will have to be privately financed, supported by non-recourse or limited recourse loans, with long-term agreements financed by the Nigerian Electricity Market, this makes it necessary for the electricity market to be functional and healthy to attract the desired local and international investment and thus translate to improve electricity supply to the average Nigerian.

The regulators (NERC) must establish a regulatory framework that ensures the financial health of utilities while incentivising operational efficiency. Tariffs must be designed such that utilities are able to recover costs with a reasonable rate of return, maintain the technical health of the electricity system, retain and expand necessary staff, expand infrastructure to meet growing and unmet demand, and accomplish other objectives. The argument above is hinged on the fact that financially healthy utilities can invest in system improvement and borrow capital from private institutions at lower interest rates, reducing debt service costs to ratepayers. Financial health also reduces uncertainty that a utility will be able to honour power purchase contracts from independent power producers.

Facilitate Private Investment

Private investments help to take the strain off utility balance sheets or Government budgets as capital expenditure and associated financing costs are avoided. To the extent that a power sector is open to private investment, and perhaps aims to increase such investment, the Regulator must create a stable investment ecosystem. The Regulator will help to facilitate power purchase or transmission use agreements, and provide private investors with certainty that those contracts will be honoured. Avoiding erratic and non-transparent decision-making helps to reduce investors' perceived regulatory risk and contributes to keeping the utilities' cost of capital low.

Protect the Interests of the Poor

As captured in section 32 (2 c) one of the roles of the NERC is to establish appropriate consumer rights and obligations regarding the provision and use of electricity services. Power sector regulators have often had statutory mandates to protect the interest of low-income customers. One such way they have fulfilled this role is in providing free or low-cost electricity to poor customers through rate balancing and weighted average—by potentially increasing rates to other customer classes. Government for their part and in some cases ameliorates the situation by providing subsidies to cushion the effect on other rate-paying classes.

Support Technical Safety and Reliability of Power System

To ensure the sustainability of the emerging market, regulators must work with other players such as the system operators, and other requisite agencies and operators, to develop and enforce standards for the safe interconnection to and operation of the power system. To this end, the Act empowers the Regulator to, amongst other things, establish technical requirements for T&D expansion and power system components, rules for interconnection of utility-scale and distributed generation (DG) systems, and standards for maintenance practices and data collection systems (section 32 (2 b, d–g) EPSR Act 2005). To efficiently carry out its roles as enunciated in the Act, the Commission/Regulator is enjoined to work collaboratively with other Government actors, in facilitating an energy-secure power sector (Jamasb and Pollitt 2008). By doing so, they should focus on creating frameworks to promote long-term security of supply for fuels used at generation facilities, minimise the frequency of fuel scarcity events, and insulate captive consumers from fuel price volatility and scarcity-related spikes by utilisation of a diverse portfolio of domestically produced energy sources as a key strategy to mitigating the risk of abrupt supply disruptions or fuel price spikes (Lee et al. 2012). The Regulator is required to assure risk-adjusted capacity expansion decisions are made (Binz et al. 2012). In addition to the roles described above, Zinaman et al. (2014) listed the following as emerging objectives which are induced by technological advancement, environmental changes, evolving social priorities, and global events. They include as follows:

- Reduce Health and Environmental Impacts of Power System Operation
- Meet Rapidly Growing Demand While Minimising Environmental Impacts and Risk
- Support Procurement of RE
- Integrate Renewable and Distributed Generation Resources to the Grid
- Incentivise Energy Efficiency, Demand Side Management, and Smart Grid Technologies

- Utilise Micro Grid/Super grid Technology
- Facilitate Consumer Participation in Power Markets
- Enhance Cybersecurity and Protect Privacy
- Manage Increased Interactions with other Sectors

We now provide brief context to some of these objectives.

Health and Environmental Impacts of Power System Operation

The environmental impact of electricity generation is significant because modern society uses large amounts of electrical power. This power is normally generated at power plants that convert some other kind of energy into electrical power. Electricity generation could have a significant impact on local and regional air quality, public health, water resources, and global carbon emission levels. Mandates to reduce environmental impacts may clash with objectives to contain costs, meet growing demand, and maintain system reliability. To meet environmental requirements, utilities may be required to invest in high-cost-mitigating technologies, such as emissions control systems or cooling technologies that lower impacts on water. Regulators are tasked with overseeing and approving these decisions and balancing environmental objectives with others in planning and capacity procurement decisions. This environmental objective when juxtaposed with the rising demand for electricity and the goal of ensuring low prices for electricity, portend further challenges for the regulators.

Support Procurement of Renewable Energy

There has been an intensified interest in accelerating RE deployment in many settings. RE penetration in Nigeria is still in its nascent stage; the only source of RE in the country is hydro-power and biomass. Wind and solar energy have only been deployed in minuscule amounts.

Regulators are tasked with designing RE incentives and regulating their integration into T&D systems. Several authors have argued that different incentives have been utilised in achieving the RE development goal such as feed-in-tariffs (Couture et al. 2010; Tongsopit and Greacen 2013), public sale (long-term) contracts (Maurer and Barroso 2011) with a reduced cost to consumers. RE is identified as one of the means of tackling the global challenges of climate change. While conventional technology costs change relatively slowly, a key challenge of RE regulation is setting tariffs in the context of declining RE technology costs. In line with the National Policy on RE and Energy Efficiency, the Commission has approved suitable windows for grid connected RE projects.

The NERC is committed to stimulating investment in RE generation in Nigeria. With a vast and mostly untapped potential in RE resources, the Commission has set a target of generating a minimum of 2000 MW of electricity from renewables by the year 2020.

- Net-metering for very small capacities (typically below 1 MW).
- Feed-in tariff for capacities up to
- 5 MW of solar,
- 10 MW of wind,
- 10 MW of biomass, and
- 30 MW of small hydro.

Integrate Renewable and Distributed Generation Resources to the Grid

To obtain a good energy mix in the electricity value chain, which, from research, is a panacea to energy sustainability, the Commission is tasked with setting rules that ensure reliable and safe operation as well as appropriate allocation of cost associated with RE and its integration across various stakeholders. In compliance with its role, the Commission has devised a model for competitive tender for capacities above the thresholds for off-grid to be procured through the Nigerian Bulk Electricity Trading (NBET) Company. In recognition of the need to address infrastructural limitations in the sector, the Commission has, in addition to facilitating financial arrangements, established in 2015 a feed-in tariff for RE based power generation in Nigeria.

Facilitate Consumer Participation in Power Markets

Research has shown the impact of consumers' participation in power markets in terms of benefits for individual consumers as well as the power system (Gelazanskas and Gamage 2014). However, this new role portends more challenges for the Regulator, as they are made to explore win-win solutions in terms of tariff, balancing demand with supply, grid stability and other related issues, to bring about optimal satisfaction to the consumers.

Enhance Cybersecurity and Protect Privacy

As more complex information and communication technology (ICT) systems are being deployed in modern grids, new cybersecurity challenges and privacy concerns are arising as well (Cárdenas and Safavi-Naini 2012). Regulators, in conjunction with other Government actors, are increasingly being asked to incorporate cybersecurity considerations into their reliability planning (Pearson 2011). Furthermore, as ICT systems generate more consumer data, regulators may be asked to issue regulations which preserve the privacy of consumers (Malashenko et al. 2012).

Manage Increased Interactions with Other Sectors

Interactions with other sectors of the economy are widening the traditional scope of power regulation. These include growing interaction with water and food systems, transport infrastructure (e.g. electric vehicles), and relationships between RE and natural gas markets (Bazilian et al. 2011; Cochran et al. 2014; Lee et al. 2012). These energy system integrations or "nexus" issues are gaining traction in international and country-level policy dialogues and are likely to be increasingly incorporated into power sector decision-making.

Sustaining a Developing Energy Sector Through Regulation

The foregoing discussion underpins the imperative of a Regulator in the power sector. Research, again, has shown how regulation is viewed as the panacea to solving market challenges in a newly privatised market such as Nigeria. This grandiloquence has increasingly been influential, leading to several nations establishing regulatory bodies to monitor and enforce the activities of the market. The electricity market varies from nation to nation, just as the regulatory map of these countries does, rendering the one-size-fits-all mantra unfeasible.

However, one characteristic feature of the regulatory body in the power sector in these countries is that two-thirds of them have juridical independence. Unlike some countries where one regulatory institution is responsible for other utilities, such as gas, water, telecom, and electricity (e.g. Belgium, Colombia, Germany, Ghana), Nigeria operates a model where there is a unique regulatory set-up for the power sector alone (e.g. the UK, France, Portugal, India, Pakistan, Argentina, Kenya, and South Africa). This implies that their mandate also differs from country to country, with responsibilities ranging from minimal activities such as price regulation as in Kenya, while in others they might have wider responsibilities such as policy making in the electricity industry, including the design of market structure and implementation of deregulation process, as in Argentina, Chile, Germany, South Africa, and the UK.

In all, one common thread that runs through them is the fact that they are deemed independent 'institutions, with a strong argument surrounding the degree of independence in practice' (Berg et al. 2000). This assertion is hinged on the fact that in many countries, regulators are faced with either direct interventions by the Government on an ad hoc basis, or continually operate under the influence of Governments in Nigeria as one of the departments of the Ministry of Power, as argued by many, despite their legally independent status. The above view was further corroborated by Dubash and Rao (2008), as being the same in India. Similar views are expressed for Kenya and Ghana (Bayliss and Fine 2007).

Critical Factors for Effective Regulatory Impact in the Nigerian Power Sector

In view of the foregoing, the impact of the Regulator in bringing about sustainability to the privatised sector in Nigeria will be predicated on three critical issues. The first is the regulatory capacity which is contingent upon the powers assigned and resources available to the regulatory institution (including funding, education, skills, and experience of personnel such as accountants, lawyers, inspectors, and engineers, along with good remuneration). Several arguments in the media portend that capacity and timely appointment/inauguration of the Regulator has had a negative impact on the market.

The governance structure of the Regulator has as much of an influence on the operations and performance of the power sector as do regulatory policies. The expiration of the term of Dr Sam Amadi led the Commission give the President Buhari Administration an opportunity to infuse new blood into electricity regulations and further the regulatory strengthening of the power sector. Section 40 of the EPSR Act mandates the President to appoint a new Chairman within three weeks from the date of the vacancy of the Chairmanship position, subject to confirmation by the National Assembly.

Furthermore, the Electric Power Sector Reform Act (EPSR) 2005, to avoid the creation of any vacuum at NERC, stipulates under Sector 35 (5) that "all appointments or re-appointments of commissioners shall be made before the expiry of their term of office in accordance with Section 34 of this Act." This unfortunately was not the case and there was a vacuum for 13 months before new Commissioners were appointed without a Chairman and a board. The impact of this vacuum on the market, according to energy experts, was astronomical. This view was corroborated by the Managing Director, Seplat Petroleum Development Company, Dr Austin Avuru, at a media brief, who was quoted by *Vanguard Newspaper* (2017) to have said:

> *Between 2015 and 2017, 18 months of regulatory lacuna disorganised the power sector reform program: created operational indiscipline, truncated*

MYTO, created a huge liquidity gap of over One Trillion Naira (NGN1 trillion). The mess is now difficult to reverse, ineffectiveness of Nigeria Electricity Regulatory Commission, NERC, fostered by a vacuum in the position of Chairman since December 2015.

Similarly, research shows that this was the second time an administrator would oversee NERC's activities before a board had been reconstituted. The first time was in 2009 when the pioneer commissioners were removed over alleged corruption, and an administrator was appointed to lead the Commission for 22 months, contrary to the provisions of EPSR Act 2005.

Furthermore, relative to criteria for appointment, Section 34 (1) of the EPSR Act provides that "subject to subsections 2 and 3 of this Section, the Commission shall consist of seven full-time Commissioners appointed by the President and subject to the confirmation by the Senate."

According to subsection (2), "in selecting potential nominees, the President shall ensure that individuals are chosen from both the public and private sectors, for their experience or professional qualifications in the following fields or areas of competence: (a) generation, transmission, system operation, distribution or marketing of electricity (b) law, accountancy, economics, finance or administration."

Subsection (3) provides that the "seven commissioners shall be appointed to reflect one commissioner per geopolitical zone and a Chairman from any zone." However, while the Chairman's tenure of office is five years, the vice chairman and other commissioners are appointed for a four-year tenure.

The EPSR Act also prescribes qualifications or areas of competence that nominees for the position of Chairman and NERC Commissioners must possess. Under Section 34 of the EPSR Act, a drawback to the appointment of capable and competent individuals as Commissioners is that each of the six geopolitical zones must produce one Commissioner. Considering that electricity regulation is still in its infancy, this requirement by the Act on political balancing, limits the pool of experienced and capable individuals that could possibly be appointed as Commissioners. Besides appointing experienced Commissioners, the technical and operational capabilities of NERC need to be further strengthened to address obvious gaps in the Commission's regulation of the power sector, which can be reactive rather than proactive. This should be an immediate priority of the new Chairman and Commissioners.

The second is the institutional environment in which regulators operate. The existence of independent and politically insulated regulatory institutions with appropriate tools and powers is not sufficient. The underwriter of the success of the private electricity market would be an independent and effective Regulator. The EPSR Act rightly established an independent Regulator. NERC is independent of Government and Market Operators. It can therefore inspire the confidence of stakeholders.

The reason for creating independent regulators is to banish fears of excessive regulatory risks, especially in countries with a history of Government interference in business operations. Such regulatory risks amount to disincentives to investment. So, to secure an attractive investment environment, Government must deliberately insulate the Regulator from pressures of Government bureaucracy.

But, in real life what is on paper as law may be very different from what is experienced in the market. Factors such as court challenges, legislative overrides, financial markets, and public interest all constrain independence, but they inject accountability. These forces make democracy work and the economy run. Effective regulation is accountable to them (Mogel 2011).

The temptations for a Government Ministry, such as the Ministry of Power, or even the Presidency, to feel compelled to interfere unduly in the decisions that the Regulator makes regarding the working of the market, are such that it may be irresistible.

To resist this temptation is a critical discipline that determines whether the nascent electricity market in Nigeria will survive. Now note this: Independence does not mean recklessness or lack of accountability. Independence means that the Regulator should be free to make the best decisions based on the evidence available to it.

The Regulator must be independent to achieve the legislative mandate granted to it by Government. But, the Regulator is accountable to political authorities and the market. Accountability is an even more important feature of an effective Regulator.

A Regulator is accountable to follow the laws establishing it and mandating its function. It is accountable to other political institutions that oversee its administrative and financial transactions. It is accountable to the ordinary jurisdiction of the court of justice, in the case of Nigeria, the Federal High Court, to review its decisions and interventions.

The effective Regulator is also accountable to operators and consumers to ensure that decisions are made after due process and through a consultative process.

In the current private electricity market, the Government needs to go through a culture change. It must realise that the market must be governed through rules made and administered by the Regulator. It must learn to focus on policymaking and use other established methods to oversee the work of the Regulator. If the Government falls into the temptation of usurping the work of the Regulator, it increases the risks of the new market and truncates the reform.

The third is the possibility of regulatory capture through which private interest groups influence the way the regulatory process is designed and implemented to appropriate utility rents to the detriment of the public interest which then leads to socially non-optimal outcomes (Stigler 1971). Arguably, regulatory independence is key to private sector confidence in the power sector and has a significant effect on the ability of Government to attract and sustain investments in the power sector.

In addition, to the critical factors discussed above, indices such as culture, politics, and social context play a significant role in general and commitments and credibility's of the contracting parties. It has been further argued that other factors such as the legal system, dispute settlement mechanisms, efficient administration of regulatory matters by public institutions, and corruption and bankruptcy procedures all influence the effectiveness of regulation. In all, research shows the degree of regulatory effectiveness differs from country to country, with several weaknesses identified in the regulatory objectives, processes, capacity, and institutional environment in developing countries (Bell 2003; Minogue and Cariño 2008; Kirkpatrick and Parker 2004a, b).

Overall, as pointed out by von Hirschhausen et al. (2004), "regulation is a repeated game between the Regulator and the regulated enterprise." Hence, the emphasis in recent literature, on regulation being the "cure for market failures after privatisation," is problematic in some respects. Notwithstanding its lofty relevance, regulatory effectiveness, however defined, requires regulatory capacity, which continues to evolve even in the developed countries and remains far from the ideal.

Having stated the above argument, it is imperative to state here that a "perfect" regulatory system is far from existent. What is sustainable, however, is the ability to make improvements and adjustments whilst adapting to internal and external changes. This process of adjustments, improvements, and adaptation can only be viable with a clear delineation of roles and relationship between Government and the Regulator as necessary to ensure reduction of investor risks and protection of consumers' interests. This can only be possible with regulatory autonomy, free from influence from external sources in its decision-making, ensuring that political oversight is not seen to impede the functioning of the Regulator. This fortunately, is possible within the Nigerian setting, where balancing regulatory autonomy/independence with sustainable financing of regulatory agencies is not difficult, since the principal source of financing for NERC is through the tariff/market funds and not through budgetary appropriations, unlike other jurisdictions, which Ogus (2005) argues makes the Regulator partly vulnerable to political influence.

Another critical and important factor that will enhance an effective regulation capable of sustaining the newly privatised market is the need for clear delineation of the policy and regulatory roles. The clear-cut roles will lead to market sustainability and will bring solution to the ever-increasing outcry of bureaucracy and 'red-tapism', which is a major hurdle to competitiveness and market sustainability. Research is full of the effect of such red tape and micro-management of the market and its forces in Nigeria. The current Commissioners are unable to carry out their regulatory roles due to over-intervention and involvement of the Ministry in the functions of the Regulator. It has been argued that this hurdle has led hugely to the failure of regulatory institutions, tools, and processes in delivering consistently "fit for purpose," user-friendly regulations and regulatory frameworks (Odion 2016).

Policy-makers have an important role to play in encouraging the electricity sector to pursue the most efficient paths to achieve energy goals. Policies need to be integrated across the power value chain to ensure that upstream fuel supply, generation assets, and T&D develop in harmony and investors are not left with fully operational but stranded assets that cannot earn a return due to lack of discipline and firm monitoring of the value chain, in turn leading to lack of customer access.

A clear-cut delineation of rules between the Regulator and the policy-makers is key to ensuring integrated and operational economic aspects, such as regulated tariffs that impact on the viability of different partici-pants in the value chain. Sustainability of the sector is hinged on a regu-lated market led by an independent and transparent Regulator who can ensure that the market is technically and financially viable across the value chain, keeping it clear of financial obstacles, benchmarking dis-tributors to reduce losses, and ensuring that the viability of generators is not threatened by fuel prices that may hinder security of supply. The integrated nature of the value chain affects not only the physical flow of electricity but also the financial flows. If there are obstacles preventing the flow of funds from customer to companies to other stakeholders (includ-ing investors) then future investment will also stall. The Nigerian power sector is seriously burdened with debt of about a trillion naira (over three billion dollars). Industry experts have attributed this to regulatory risk, as discussed earlier in the chapter (Madu 2017).

A well-structured regulatory framework is important in ensuring a level playing field for private players in the NESI. By so doing, transpar-ency in overall industry governance and clear separation between policy-makers and regulators can be ensured or guaranteed. This will further lead to a more liberalised and commercialised market where large com-mercial and industrial customers, through the corridor of eligibility, can purchase power from the generation companies.

Sustaining the gains of the reform in Nigeria entails the ability to meet the electricity demands of residential, commercial, and industrial con-sumers, leading to an unprecedented level of investment in fuel supply, centralised and DG, and networks across the whole value chain. The combined effects of growing demand for electricity and rising capital intensity will mean increased investment and expansion, with renewables adequately included in the mix.

Conclusion

It is well known that the recently reformed electricity sector is undergoing serious and insightful changes, leading to significant transformation with rapidly evolving technologies, declining costs, shifting regulatory landscapes and, in fast-growing economies, rising demand (Ogus 2005). The success of the Nigerian power sector portends success for several other sectors, which will lead to unlocking economic potential and improving living standards. The role of the Regulator is complex. Ensuring reliable service at reasonable cost involves balancing the interests of utility investors, energy consumers, and the entire economy. The Regulator should ensure that both suppliers and consumers uphold their contractual and licence obligations whilst ensuring that the utility has the obligation (via licensing) to provide services under the approved tariffs and quality standards. Consumers have an obligation to pay for services supplied to ensure the financial viability of the sector.

Sustainability of the electricity market in Nigeria and any other jurisdictions must ensure a market setting which aims to ensure the effective protection of the property rights of investors, and provide a framework of known legal rules. It has been argued that the act of copying and pasting models which succeeded in other jurisdictions without adequate indigenisation is a mission in futility. This is so as several exigencies which are peculiar and fundamental to the success or failure of such models in such countries are ignored. One of the commonly neglected reasons for this is fewer resources being available in developing countries. For instance, evolving economies such as Nigeria is grappled with technological and capacity deficiency predominantly in the regulatory commission. The Regulator is unable to adapt effectively to the market conditions. Also important is capacity; Nigeria as opposed to using sound and relevant capacity measures, predominantly focuses on the so-called federal character of ensuring every region in the six geopolitical regions are represented by employing and appointing persons who are at best round pegs in square holes.

Yet another, shortcoming is the lack clarity of the relationship between Government and Regulator, notwithstanding the clear delineations of the roles of the regulatory and policy roles, there have been several cases of conflicting roles. To date, Nigeria lacks the political will to allow regulators to "do their jobs" (Odion 2016).

Research has shown that privatisation can only work effectively in a well-structured and competitive settings or market structures. To achieve this, there is need for effective regulation capable of encouraging and bringing about competition in the market amongst the operators. The current situation, where, due to the factors enunciated above, the Regulator is unable to carry out its mandate and instead created so much uncertainty for the operators who have little or no leverage in making the needed business decisions, is counterproductive for all parties (Kemal 1996). Disputably, there is no "perfect" regulatory system. However, considering the relevant principles, as well as the implementation of industries'—international or local—best practices, continuous improvements and adjustments are necessary as the system adapts to internal and external changes (Kemal 1996). Also, clarity of the relationship between Government and the Regulator is crucial to a good reduction of investor risks and protection of consumers' interests whilst ensuring the three basic principles of regulation: independence, transparency, and investor/consumer protection.

References

Amadi, Sam. 2014. *The Regulatory Powers of NERC under the Extant Laws in the NESI*. A Paper Delivered at the 3rd Seminar for Judges on Regulation in the Electric Power Sector Organised by The Nigerian Electricity Regulatory Commission, Uyo, Akwa Ibom State.

Awosope, Claudius O. A. 2014. *Nigeria Electricity Industry: Issues, Challenges and Solutions*. Covenant University 38th Public Lecture.

Balducci, Patrick J., Joseph M. Roop, Lawrence A. Schienbein, John G. DeSteese, and Mark R. Weimar. 2002. *Electric Power Interruption Cost Estimates for Individual Industries, Sectors, and US Economy*, No. PNNL-13797. Richland: Pacific Northwest National Laboratory (PNNL).

Baumol, William J. 1996. Rules for Beneficial Privatisation: Practical Implications of Economic Analysis. *Islamic Economic Studies* 3 (2): 1–32.

Bayliss, Kate, and Ben Fine, eds. 2007. *Privatisation and Alternative Public-Sector Reform in Sub-Saharan Africa: Delivering on Electricity and Water*. Springer International Publishing AG.

Bazilian, Morgan, Holger Rogner, Mark Howells, Sebastian Hermann, Douglas Arent, Dolf Gielen, Pasquale Steduto, et al. 2011. Considering the Energy, Water and Food Nexus: Towards an Integrated Modelling Approach. *Energy Policy* 39 (12): 7896–7906.

Bazilian, Morgan, Mackay Miller, Reid Detchon, Michael Liebreich, William Blyth, Matthew Futch, Vijay Modi, et al. 2013. Accelerating the Global Transformation to 21st Century Power Systems. *The Electricity Journal* 26 (6): 39–51.

Bell, Matthew. 2003. Regulation in Developing Countries is Different: Avoiding Negotiation, Renegotiation and Frustration. *Energy Policy* 31 (4): 299–305.

Berg, Sanford V., Ali Nawaz Memon, and Rama Skelton. 2000. *Designing an Independent Regulatory Commission*. Public Utility Research Centre Working Paper, University of Florida.

Besant-Jones, John E. 2006. *Reforming Power Markets in Developing Countries: What Have We Learned?* Washington, DC: World Bank.

Binz, Ron, Richard Sedano, Denise Furey, and Dan Mullen. 2012. Practicing Risk-Aware Electricity Regulation. *CERES [Internet]. April.* Ceres.org. http://www.ceres.org/resources/reports/practicing-risk-aware-electricity-regulation.

Cárdenas, Alvaro A., and Reihaneh Safavi-Naini. 2012. Chapter 25: Security and Privacy in the Smart Grid. In *Handbook on Securing Cyber-Physical Critical Infrastructure*, ed. Saial K. Das, Krishna Kant, and Nan Zhang. Boston: Elsevier/Morgan Kaufmann.

Cochran, Jaquelin, Owen Zinaman, Jeffrey Logan, and Doug Arent. 2014. *Exploring the Potential Business Case for Synergies Between Natural Gas and Renewable Energy*, No. NREL/TP-6A50-60052. Golden: National Renewable Energy Laboratory (NREL).

Couture, Toby D., Karlynn Cory, Claire Kreycik, and Emily Williams. 2010. *Policymaker's Guide to Feed-in Tariff Policy Design*. No. NREL/TP-6A2-44849. Golden: National Renewable Energy Lab (NREL). Available at: http://www.nrel.gov/docs/fy10osti/44849.pdf. Accessed July 2017.

Dubash, Navroz K., and D. Narasimha Rao. 2008. Regulatory Practice and Politics: Lessons from Independent Regulation in Indian Electricity. *Utilities Policy* 16 (4): 321–331.

Gelazanskas, Linas, and Kelum A.A. Gamage. 2014. Demand Side Management in Smart Grid: A Review and Proposals for Future Direction. *Sustainable Cities and Society* 11: 22–30.

Jamasb, Tooraj, and Michael Pollitt. 2008. Security of Supply and Regulation of Energy Networks. *Energy Policy* 36 (12): 4584–4589.

Kemal, Abdul Razzaq. 1996. Why Regulate a Privatised Firm? *The Pakistan Development Review* 35: 649–656.

Kirkpatrick, Colin, and David Parker. 2004a. *Infrastructure Regulation: Models for Developing Asia*, No. 60. ADBI Research Paper Series. Tokyo: Asian Development Bank Institute.

———. 2004b. Regulatory Impact Assessment and Regulatory Governance in Developing Countries. *Public Administration and Development* 24 (4): 333–344.

Lee, April, Owen Zinaman, Jeffrey Logan, Morgan Bazilian, Douglas Arent, and Robin L. Newmark. 2012. Interactions, Complementarities and Tensions at the Nexus of Natural Gas and Renewable Energy. *The Electricity Journal* 25 (10): 38–48.

Madu, Belije. 2016. *Show Me the Money: The Nigeria Electricity Supply Conundrum*. Available at https://opinion.premiumtimesng.com/2016/11/02/175607/. Accessed on 14 Oct 2017.

———. 2017. *Solve Electricity; Solve Nigeria's Economy Issues*. Available at https://www.pmnewsnigeria.com/2017/05/22/solve-electricity-solve-nigerias-economy-issues/. Accessed 14 Aug 2017.

Malashenko, Elizaveta, Chris Villarreal, and J. David Erickson. 2012. *Cybersecurity and the Evolving Role of State Regulation: How It Impacts the California Public Utilities Commission*. San Francisco: California Public Utilities Commission. Available at: http://www.cpuc.ca.gov/NR/rdonlyres/D77BA276-E88A-4C82-AFD2-FC3D3C76A9FC/0/TheEvolvingRoleofStateRegulationinCybersecurity9252012FINAL.pdf. Accessed July 2017.

Maurer, Luiz, and Luiz A. Barroso. 2011. *Electricity Auctions: An Overview of Efficient Practices*. World Bank Publications: Latin America.

Minogue, Martin, and Ledivina Cariño, eds. 2008. *Regulatory Governance in Developing Countries*. Cheltenham: Edward Elgar Publishing.

Mogel, William A. 2011. Preside or Lead? The Attributes and Actions of Effective Regulators. *Energy LJ* 32: 627.

Odion, Omonfoman. 2016. *Ensuring Regulatory Independence in the Power Sector*. Available at https://opinion.premiumtimesng.com/2016/01/11/ensuring-regulatory-independence-in-the-power-sector-by-odion-omonfoman/. Accessed 14 Aug 2017.

Ogus, Anthony. 2005. *Towards Appropriate Institutional Arrangements for Regulation in Less Developed Countries*. Centre for Regulation and Competition, University of Manchester, Working Paper Series, Paper No. 119.

Pearson, Ivan L.G. 2011. Smart Grid Cyber Security for Europe. *Energy Policy* 39 (9): 5211–5218.

Sanghvi, Arun P. 1982. Economic Costs of Electricity Supply Interruptions: US and Foreign Experience. *Energy Economics* 4 (3): 180–198.

Stigler, George J. 1971. The Theory of Economic Regulation. *The Bell Journal of Economics and Management Science* 2: 3–21.

Tongsopit, Sopitsuda, and Chris Greacen. 2013. An Assessment of Thailand's Feed-in Tariff Program. *Renewable Energy* 60: 439–445.

Vanguard Newspaper. 2017. Economy Staved of N534bn Power Sector Inefficiency in 2016- Avuru at the 7th Emmanuel Egbogah Legacy Lecture Series. Themed: Nigeria Petroleum Industry: The Dawn of a New Era. Available at: https://www.vanguardngr.com/2017/09/economy-staved-n534bn-power-sector-inefficiency-2016-avuru/.

Von Hirschhausen, Christian, Thorsten Beckers, and Andreas Brenck. 2004. Infrastructure Regulation and Investment for the Long-Term—An Introduction. *Utilities Policy* 12 (4): 203–210.

Zinaman, Owen, Mackay Miller, and Morgan Bazilian. 2014. *Evolving Role of the Power Sector Regulator: A Clean Energy Regulators Initiative Report*, No. NREL/TP-6A20-61570. Golden: National Renewable Energy Laboratory (NREL).

Part II

Management and Cross-Cutting Issues

4

Stakeholder Engagement and the Sustainable Environmental Management of Oil-Contaminated Sites in Nigeria

George Prpich, Kabari Sam, and Frédéric Coulon

Introduction

The extractive industries in Africa are experiencing a period of rapid growth and development, but progress comes at a cost both to the environment and to the society. In Nigeria, years of neglect and mismanagement have resulted in vast tracts of land being contaminated by

G. Prpich (✉)
Department of Chemical Engineering, University of Virginia,
Charlottesville, VA, USA
e-mail: prpich@virginia.edu

K. Sam
Environment and Conservation Unit, Centre for Environment, Human Rights
and Development, Legacy Centre, Port Harcourt, Nigeria
e-mail: kabari.sam@cehrd.org.ng

F. Coulon
School of Water, Energy, and Environment, Cranfield University, Bedford, UK
e-mail: f.coulon@cranfield.ac.uk

© The Author(s) 2019   **75**
S. Adesola, F. Brennan (eds.), *Energy in Africa*,
https://doi.org/10.1007/978-3-319-91301-8_4

hydrocarbon pollution. Efforts to reverse these impacts and to prevent further harm have proved to be ineffective. Alternative methods that can address the limitations of regulation and integrate the values and perspectives of the multitudes of stakeholders who live with the pollution every day are needed to manage the environment.

Stakeholder engagement is a methodology that can collate the viewpoints of different stakeholders affected by a business's objectives. Stakeholder engagement is used by businesses to build trust, to promote transparency, and to gain a social licence to operate. The literature has focused primarily on the role of the stakeholder relative to the development of private sector strategic plans. This chapter aims to introduce the concept of stakeholder engagement and discuss how it can be used to assist the environmental management of oil-contaminated land in Nigeria. Current challenges and opportunities for stakeholder engagement to support policy development are given and illustrated through a case study.

What Is Stakeholder Engagement?

A stakeholder can be defined as any individual, or group of individuals, that might be affected by, or cause an effect on, the ability of an organisation to achieve their objectives (Cundy et al. 2013; Geaves and Penning-Rowsell 2016). In this context, a stakeholder can be any of the following: shareholders, employees, customers, suppliers, lenders, collaborators, and society.

Stakeholder engagement is a process by which an organisation informs, consults, involves, collaborates with, and empowers groups and individuals affected by a decision, operation, or policy (Rowe and Frewer 2005; Ramirez-Andreotta et al. 2014; Benson et al. 2016). Stakeholder engagement emerged as a meaningful area of research in 1984, when Ed Freeman's seminal book *Strategic Management: A Stakeholder Approach* (Freeman 2010) popularised a practical pathway for implementing the concept. His work brought into focus the concept of the Stakeholder Approach and suggested a pragmatic framework for implementing and delivering on the concept (Fig. 4.1).

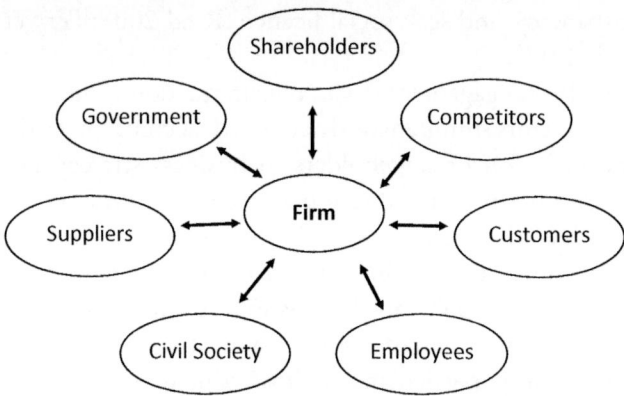

Fig. 4.1 Stakeholder approach. (Adapted from Freeman 2010)

The Stakeholder Approach was conceptualised as a holistic methodology used by organisations to understand the concerns of shareholders, employees, customers, suppliers, lenders, and society. These concerns might be economic, social, environmental, or political in nature, and might relate to issues of fairness, trust, inclusion, or control. The Stakeholder Approach was developed to enable the organisation to learn and understand from the concerns of the stakeholders; information that could provide insight to help an organisation develop business strategies and objectives that stakeholders could support. By understanding these concerns and interactions, it was believed that a mutual benefit could be generated. Fundamental to this approach was the idea that stakeholders should be viewed as people, each with their own unique set of values and aims (Slinger 1999).

Evidence suggests that when the various views of different stakeholders are considered, the overall quality of a decision is enhanced (Garmendia and Stagl 2010; Cundy et al. 2013; Sardinha et al. 2013). Stakeholder engagement has become an integral element of decision-making both in the public and in the private sectors. It is used to inform, consult, and create dialogue among stakeholders and therefore empower interested parties to participate in the decision processes that might impact on them (Reed 2008). For complex issues, stakeholder engagement is also used to identify gaps in knowledge, reveal perceptions of risk, build trust, promote transparency, and seek social licence (Reed 2008; Péry et al. 2013; Prpich et al. 2011).

As stakeholder engagement began to gain traction, a number of proponents and detractors simultaneously emerged. A common criticism levied against the inclusion of stakeholders in business strategy development was the presumption that the sole purpose of the firm was to act in its own calculative self-interest to maximise profits for the benefit of the shareholders. When the self-interests of the firm differed from those of the stakeholders, it would not be possible to maximise the benefits of everyone. Critics thus viewed the inclusion of stakeholder interests as a logical impossibility, particularly if those benefits were multidirectional (Slinger 1999).

Proponents of the stakeholder approach argued that calculated self-interest missed the point of the concept. The aim was not solely to maximise benefits to all stakeholders, but rather is a gesture of moral commitment to all stakeholders whereby the firm provides an opportunity for stakeholders to receive benefit by the promotion of business practices that enable the realisation of opportunities. Other arguments against stakeholder engagement—for example stakeholders do not want to be involved in business decisions; knowledge about how a mineral is extracted will make stakeholders oppose the process; and the entire activity is too resource-intensive—have since been dismissed. Involving stakeholders in the decision-making process through stakeholder engagement is now accepted to be an effective means for building mutually beneficial relationships (Devin and Lane 2014).

Stakeholder engagement needs to be more than an instrumental management tool (Wheeler et al. 2002). Wheeler suggests that engagement should be integrated within a business's strategy and that appropriate resources (e.g. time) must be committed to address the complexity of socio-environmental problems. To be effective, engagement with the community must occur at all levels of the business. The firm must commit to listening, understanding, and acting upon information they receive, even if that information pertains to perceived risks (e.g. environmental or societal risk, organisational or government behaviour). Any point of contention or misunderstanding could lead to conflict. Cursory stakeholder engagement is therefore not sufficient for understanding and communicating the impacts of complex social and environmental issues (Mzembe 2016). Good practice in stakeholder engagement means that firms are committed to the process and that engagement should continue to occur throughout the development process, and the lifetime of an operation.

Corporate Social Responsibility

Corporate Social Responsibility (CSR) broadly encompasses a firm's efforts to address the social and environmental issues which result from their operations that extend beyond the efforts that comprise their normal pursuit of profits (Vogel 2007). More recently, CSR has come to be defined as the "triple bottom line" of financial, environmental, and social responsibility and is rapidly becoming the norm regarding sustainable business practices (Jha and Cox 2015).

CSR is likely to have begun as a voluntary, charitable pursuit; however, the adoption of CSR by the extractive industries could also be a response to the sector's history of various economic, environmental, and social issues. Though the sector provides employment and creates wealth, it has also been plagued by environmental disasters, land disturbances, depletion of non-renewable resources, and threats to the health and safety of workers and citizens (Azapagic 2004). In this context, engagement with stakeholders could be an invaluable mechanism for organisations to communicate the risks and benefits associated with various extraction practices. Recent empirical evidence supports this idea and shows that CSR can be a mechanism for increasing profits, engendering trust, and securing a social licence to pursue business (Henisz et al. 2014; Jha and Cox 2015; Wang and Sarkis 2017).

CSR is underpinned by a commitment to recognise, internalise, and respond to societal concerns and expectations by means of stakeholder engagement. Its use has become so commonplace that many regions now view stakeholder engagement as a regulatory and political imperative, without which trust and access might not be granted (Moomen and Dewan 2017). A review of CSR reports for mining companies illustrates the commitment that organisations have made to develop good community relationships nurtured through stakeholder engagement (Jenkins and Yakovleva 2006). Firms that increase their commitment to stakeholder support have been shown to be more likely to increase the financial valuation of their firms (Henisz et al. 2014).

In the next section, we ask the questions: How might Government organisations benefit from these stakeholder concepts, and how might stakeholder engagement be used to support the transfer of environmental policy to manage the impacts associated with petroleum hydrocarbon extraction?

Can the Government Operate Like a Firm?

Stakeholder approaches are well established at the firm level and are solidly embedded within the CSR framework, yet the practice of stakeholder engagement within government institutions is not as well established. Flak and Rose (2005) conceptualise the institution of government as a firm, where the management of relationships and interests of societal stakeholders is paramount. It is possible to imagine a government using stakeholder engagement to similar ends as a firm, given that democratic political models used by governments are motivated to balance the legitimate competing interests in society (Flak and Rose 2005). Though governments are not motivated to maximise the profits of shareholders, such as in a firm, there is a strong desire to operate as a business insofar as optimising budgets and managing change (Scholl 2004). Though the concept of stakeholder engagement remains the same across public and private sectors, the tools and instruments for implementing stakeholder engagement might differ (Bingham et al. 2005).

Specific to the environment, governments can use stakeholder engagement to understand and communicate the impacts (past, present, and future) of petroleum hydrocarbon industrial activities, and to secure a social licence to support development (Moomen and Dewan 2017). Stakeholder engagement enables social legitimacy and, when accompanied by social programmes to enhance the social welfare of affected populations, governments could increase their capacity to intervene and support side-lined communities (Bawole 2013). Too often, stakeholder engagement is bundled into the Environmental Impact Assessment (EIA) process where it becomes little more than a cosmetic tick box exercise done for the purpose of meeting legal requirements (Bawole 2013).

In the pursuit of a stakeholder engagement framework, government should avoid the adoption of a single, overarching strategy for community involvement, which may not be helpful in regions such as Nigeria where communities' structures and situations differ greatly. Building engagement within communities that struggle with weak local government entities and a large, illiterate local population requires tailored communication responses (Bawole 2013). In the following case study, we outline an approach intended to assist governments to engage with different stakeholder communities affected by the impacts of oil extraction in the Niger Delta region. The case study seeks to understand how different stakeholder communities value certain socio-cultural, economic, and environmental values, how decisions to prioritise these values are made, and how engagement might inform contaminated land management policy.

Case Study: Stakeholder Engagement to Support Contaminated Land Management Policy in the Niger Delta, Nigeria

A Legacy of Crude Hydrocarbon Pollution

Situated in southern Nigeria, the Niger Delta sits at the apex of the Gulf of Guinea and encompasses an area of 112,110 km². Divided into nine states, the region is home to a population of approximately 31 million people (NDDC 2006). Much of the population relies on the land and natural resources for their livelihood, which largely comprises subsistence farming and fishing (Chinweze et al. 2012).

Oil was discovered in the region in 1956 and vast reserves have been exploited consistently since. Nigeria has generated considerable wealth from this resource and oil exports were valued at US $89 billion per annum (OPEC 2015). These values have contributed up to 35% of

Nigeria's gross domestic product, and over 90% of its foreign exchange wealth (OPEC 2015). Yet despite this boom in petroleum wealth, Nigeria remains a distinctly poor nation that ranks 156th out of 216 countries in terms of Gross National Income and Purchasing Power (World Bank 2016).

The Niger Delta region has been the hub for oil extraction and processing in Nigeria for the past 50 years (OPEC 2015). Over this period, oil spills caused by engineering failure, oil theft, pipeline vandalism, and natural factors have resulted in considerable land contamination (Fentiman and Zabbey 2015; Kadafa 2012) that has impacted on human health, groundwater, soil functionality, and ecological systems (Pegg and Zabbey 2013; UNEP 2011). For example, in 2008 a major pipeline failure in Bodo City led to an oil spill that affected over 69,000 households (Pegg and Zabbey 2013). In addition to these larger spills, smaller spills occur frequently, adding to the cumulative pollution in the environment (Nwilo and Badejo 2006).

Cost to clean up watercourses, groundwater, and soil in the region was estimated to range between US $500 million and US $1 billion (UNEP 2011). Although the scale of land contamination in the region is difficult to quantify, over 2000 sites that require remediation were estimated to exist as of 2008 (Ite et al. 2013), and confirmed in 2011 by the United Nations Environment Programme (UNEP) (UNEP 2011). Despite widespread crude oil contamination there is no evidence to date to indicate that clean-up has commenced in the region.

A Contaminated Land Management Policy That Needs to Be Strengthened

Policy to manage contaminated land has developed over the last five decades and can be classified into three distinct periods (Fig. 4.2): no legislation, non-specific legislation, and specific legislation. Prior to 1968, regulations were limited and emergent—for example Oil and Pipeline Act, 1956; Mineral Oils Regulation, 1963; Petroleum Regulations, 1967; and Oil in Navigable Water Decree, 1968—addressing concerns about licensing, safety, and transportation. These efforts were generally

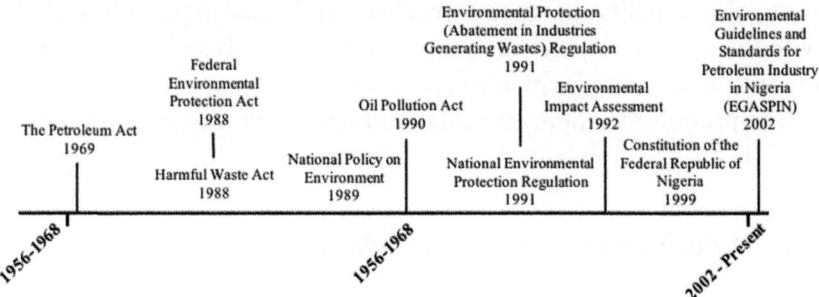

Fig. 4.2 Contaminated land policy development from 1956 to the present. (1956–1968—period of no legislation; 1969–2001—period of non-specific legislation; 2002–present—period of specific legislation). (Source: Sam et al. 2017b)

considered to be reactive and fragmented, and failed to manage contaminated land (Ite et al. 2013).

Oil production, and oil contamination incidents, increased after 1968, which elicited a response from the Nigerian authorities in the form of the Petroleum Act (1969). An overarching legislation, the Petroleum Act aimed to prevent pollution in water, air, and soil and was complemented by the Harmful Waste Act 1988 and the Environmental Impact Assessment Act 1992 in later years (Ajayi and Ikporukpo 2005).

In the late 1980s, continuing contamination events in the Niger Delta led to community protests that forced the Government to act, resulting in the development of the Environmental Guidelines and Standards for the Petroleum Industry in Nigeria (EGASPIN). Published in 1991 and enacted in 2002, the EGASPIN provided specific regulations for the management of contaminated land, which remain the basis for environmental management in Nigeria to this day.

It is acknowledged that Nigerian environmental governance has been ineffective and reasons for this include: lack of adequately trained and experienced personnel, lack of expertise, inappropriate screening values, and misaligned socio-economic expectations (Ajayi and Ikporukpo 2005; Eneh 2011). Ineffectiveness might also stem from Nigeria's approach to policy development, which is to adopt/transfer policies from other countries (e.g. the USA) (Sam et al. 2017a, b). Policy transfer is a common practice for governments that lack the technical capabilities or expertise to develop a policy on their own (Ajayi and Ikporukpo 2005). A flaw with this approach is that the adopted policy is likely to lack contextualisation: socio-cultural, environmental, economic, or otherwise, and this will impact on the appropriateness and efficacy of the policy to serve its purpose.

The Stakeholder Engagement Method

The stakeholder engagement framework used (Fig. 4.3) was modified to overcome potential barriers related to communication style, language, and understanding. Cultural preferences include contact and discussion (Idemudia 2014), therefore engagement was designed to incorporate face-to-face interviews and workshops. Workshops conducted with community members made use of postcards to graphically depict social values. Participants felt that a combination of graphics and text would overcome the challenges of language and comprehension (Zhao et al. 2016). Similar data obtained from experts and industry were collected via phone interviews, reflecting the needs of those stakeholders. Open ended questions were combined with structured prioritisation exercises to identify important issues as well as capture information about perceived risks held by the participants.

The developmental stage in the stakeholder engagement was underpinned by three activities: preliminary planning, development of values, and organisation and validation of values.

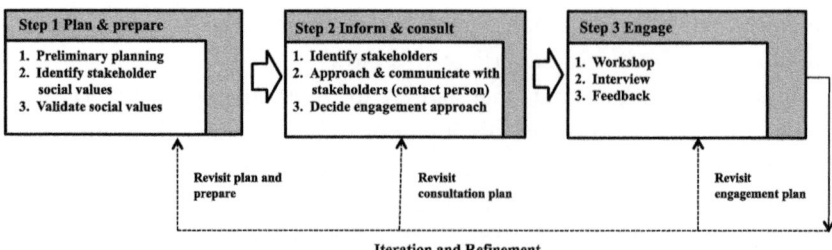

Fig. 4.3 Stakeholder engagement framework for collecting information about socio-cultural values relative to petroleum hydrocarbon pollution and contaminated land policy in Ogoniland, Nigeria. (Source: Sam et al. 2017a)

(1) *Preliminary planning:* At this stage, it was defined who should be engaged, how they should be engaged, what they would engage with, and to what extent they would be engaged. Objectives were clearly stated and resources to conduct the engagement were assessed (Cundy et al. 2013; Rangarajan et al. 2013). Stakeholders for this study were selected from areas highly impacted on by petroleum hydrocarbon pollution (Kadafa 2012). Four categories of stakeholders were identified: community members, experts, regulator, and operators (Idemudia 2014; Kadafa 2012; UNEP 2011). Stakeholders were initially identified from previous reports, for example the environmental assessment of Ogoniland (UNEP 2011), and peer-reviewed literature. Community participants were selected through a process led by community leaders.

(2) *Development of a list of values:* Socio-cultural, economic, and environmental values were determined and defined relative to the region prior to the workshop using literature searches. Development of a list of social values where socio-cultural, economic and environmental issues were identified via a critical review of the academic databases (e.g. Science Direct, Scopus) and online databases (e.g. Google Scholar) using key phrases and words, such as values, impacts, oil spills, land contamination, socio-economic and environmental impacts, stakeholder values, stakeholder concerns, contaminated land concerns, Niger Delta, Nigeria. Values represented the perceived concerns of the stakeholder, and although the provided list was not exhaustive, it did provide stakeholders with a starting point for richer discussion.

(3) *Organisation and validation of the identified values:* Before values were used in the workshop, they were validated by unofficial discourse with contaminated land experts from Nigeria. This was done to ensure that the identified values represented the actual values of stakeholders in the study area. This review provided a layer of feedback that helped to accurately contextualise the issues (Table 4.1).

The Inform and Consult stage was used to identify participants and to communicate with them the intention and purpose of the engagement process. Stakeholders might include community members, contaminated

Table 4.1 Stakeholder values as identified from literature and validated by experts

Values	Elements	Description
Socio-cultural	Communal crisis	Communal crisis refers to a crisis that exists between communities, oil companies, and government.
	Cultural places	Cultural places include places of worship and cemeteries.
	Family and household	Children, parents, and relatives.
Environmental	Drinking water quality	The water used to provide drinking water to communities.
	Loss of biodiversity	Loss of variety of flora and fauna in the local environment.
	Resource conservation	How you use, allocate, and protect your natural resources such as fish and mangrove habitats.
	Soil quality for agriculture	Maintenance of soil quality to enable agriculture for nutritional and economic value.
Economic	Food and local supply chain: farming and fishing	Sources of local food supply such as farming and fishing, and nutrition.
	Legacy for future generations	Natural resources you wish to transfer to your grandchildren are in decline.
	Human health/ wellbeing	Health and wellbeing (sickness and diseases).
	Financial issues/ income security	Financial health, the ability to sustain an income.
	Reputation	The reputation of your community or institution.
	Collaboration/ co-existence	Collaboration and cooperation among operators, regulators, community members, and government.

Source: Sam et al. (2017a)

land experts, regulators, and operators from, or working within, oil-impacted regions. Communication with stakeholders requires an understanding of preference and technical capability. Whereas contaminated land experts, regulators, and operators might prefer, and have the capability, to communicate by email or telephone, community participation is more likely to be face-to-face communication, and knowledge about events will rely on effective word of mouth or a town crier (Noy 2008; Rizzo et al. 2015).

The Engage stage describes the approach used to interact with participants. Sam et al. (2017a) conducted workshops for community members ($n = 35$), and telephone/in-person interviews for operators ($n = 7$), regulators ($n = 8$), and experts ($n = 6$). To enable comparison between groups, it was important to base engagement on a common framework of questions. For example, probing questions can be used to explore participants' depth of knowledge and to reveal perceptions of risk. Data analysis depends on the type of data collected. Qualitative data can be analysed using, for example, thematic content analysis methodology (Krippendorff 2012), whereas statistical analysis can be used to analyse quantitative data.

What Do Participants Value?

Stakeholder engagement provides insight into the perspectives, values, and knowledge held by different stakeholders. A key finding from Sam et al. (2017a) was that the different stakeholder communities prioritised values similarly. All stakeholder groups shared an acute awareness of the extent and impact of hydrocarbon pollution in the region, as well as the effect that pollution had on the health and livelihood of local communities (Fig. 4.4).

Drinking Water

Access to safe drinking water was the highest ranked value by all stakeholders. This result was not unexpected because the majority of the local population lacks access to safe drinking water (Etim et al. 2013). Community members commented that contaminated drinking water presented a daily challenge, caused sickness, and was unavoidable. Community members unable to access clean drinking water are forced to buy bottled water, which is expensive and can sometimes be unsafe for consumption (Akpabio et al. 2015). Without access to safe drinking

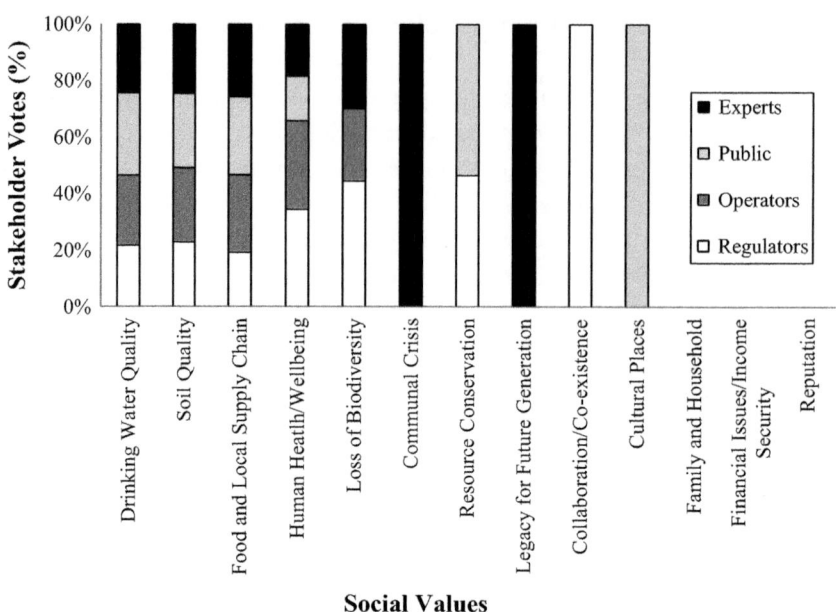

Fig. 4.4 Stakeholder voting preferences as a percentage of total votes cast for each social value presented in the study. (Values are ordered according to rank of importance from left to right. The first four social values: Drinking Water Quality, Soil Quality, Food and Local Supply Chain, and Human Health/Wellbeing, were similarly ranked for all stakeholder communities). (Source: Sam et al. 2017a)

water a stakeholder's health might deteriorate, their ability to create economic wealth might diminish, and frustration and desperation might become established. If an individual begins to feel that control has been lost they might seek out the only recourse that they believe is available to them (e.g. conflict, sabotage), which might further exacerbate the problem that initiated the cycle (Fentiman and Zabbey 2015). Operators and regulators are aware of the water contamination issue, yet have been unable to remediate the problem. Although Nigeria possesses a national water policy (FGN 2004), it has not achieved its goals of providing equitable and safe water resources due to weak enforcement and implementation (Nwankwoala 2014). Weak enforcement and implementation are central tenets of environmental policy in Nigeria. Transferring a good policy to overcome a bad policy will not remedy this issue, but integration of contaminated land management policy that works in concert with a national water policy to meet societal need might begin to address the complexity of the problem.

Soil Quality

The economy of the Niger Delta region is reliant on agriculture. Soil impacted on by hydrocarbon contamination suffers from reduced quality (Okeke and Okpala 2014), which has translated into lower agricultural yields (Oyebamiji and Mba 2013). As with drinking water, there exists an intrinsic link between soil quality, agriculture, and livelihood. Operators acknowledge the hardship in the area and regulators are frustrated by their inability to help local communities restore their agricultural vibrancy, even after remediation has occurred. It is clear that the contaminated land policy has failed to address the issue of contaminated land, whether due to inadequate generic soil standards, poor enforcement, a lack of funding, expertise, or institutional coherency (Ajayi and Ikporukpo 2005; Sam et al. 2017b). Findings suggest that policy should integrate a method to prioritise the treatment of those soils that are vital for agricultural production. Prioritisation targets sites that pose the greatest harm to local populations, and provides a mechanism for efficient allocation of limited resources (Sam et al. 2017c). Nigeria would also benefit from prompt responses to new spills. Mechanisms for reporting spills in the UK, the USA, and Canada could serve as exemplars, but these approaches are expensive. More economically relevant lessons could be gleaned from nearby countries, such as Cameroon (Forton et al. 2012). If Nigeria were to adopt the reporting protocols of wealthier nations, they could reduce costs by integrating a stakeholder approach such that citizens become the mechanism for reporting and identifying spills that inform prioritisation (e.g. see Sam et al. 2017c).

Food Supply

The health of the local population has suffered due to the hydrocarbon pollution that has affected the local food chain (Nriagu et al. 2016). Community members know that oil-contaminated food will make them ill, but they must continue to eat fish from contaminated waters and crops from contaminated lands as a means of survival. Similarly, farmers and fishers continue to practice their operations in the presence of hydrocarbon contamination because they have no viable alternatives to sustain themselves. Regulators comment that a lack of timely spill response prevents them from stopping the public's consumption of contaminated food. Time to response could be improved through local involvement to identify and report spill incidents, similar to practices in the USA (CERCLA 2002), but adoption of this practice will require broad public support and the development of trust, both of which can be generated through stakeholder engagement.

Opportunities for Stakeholder Engagement to Support Policy Development

The basis by which stakeholders assessed values differed, despite a shared prioritisation of the issues. Community members expressed concerns about the effects of contamination based on the impacts that pollution had on individual health and livelihood. Regulators expressed their concerns relative to how a value might prevent their organisation from meeting a strategic goal. Stakeholder engagement conducted across stakeholder groups could be used to share and communicate these differences, which might lead to improvements in trust, understanding, comprehension, and shared decision-making (Snape et al. 2014). Stakeholder engagement also revealed valuable information about stakeholder perceptions of risk, which can be as salient as the risk itself. Changing perceptions is an immense challenge and stakeholder engagement provides a means for Government to learn and comprehend the base of those perceptions, and to raise awareness of the issues. Public awareness, generated through education, communication, and understanding, enables the public to actively participate in the management of contaminated land. When public awareness about the impacts of hydrocarbon pollution in a region are either misunderstood, or ignored, feelings of frustration, desperation, inequality and loss of control can emerge potentially leading to an individual committing acts that perpetuate pollution, for example pipeline vandalism (Nwilo and Badejo 2005). In contrast, public awareness in countries such as the USA and UK is high, largely due to media coverage and engagement during public consultations. Stakeholder engagement is a relatively inexpensive option for Governments to raise awareness, and educate residents, land developers, and the public about the issues associated with hydrocarbon pollution and land contamination.

Resource availability is a focal challenge for Nigerian policy makers. In the event of a contamination event in the UK or the USA, communities are granted the resources necessary to relocate, remediate, and rebuild affected areas. Those in the Niger Delta region are not afforded this necessity (Fentiman and Zabbey 2015). As an example, BP was penalised approximately US $20 billion in damages to local and Federal Governments shortly after the Deep-Water Horizon accident, whereas only US $1 billion has been allocated to the clean-up of the Niger Delta, a contamination catastrophe that is decades in the making and whose impacts affect a far greater number of people. Herein lies the challenge for Nigeria and its intent to develop an improved contaminated land management policy through policy transfer. When the conditions for which a policy was developed differ from the conditions for which it will be applied, it is not reasonable to expect the same degree of efficacy and success without some effort to contextualise.

Once a policy response has been adopted, stakeholder engagement activities must continue to ensure final delivery of that policy. For example, following the United Nations Environment Programme report on hydrocarbon pollution in 2011, the Nigerian government launched the Hydrocarbon Pollution Remediation Project (HYPREP) to implement the recommendations laid out in the report. HYPREP's responsibility was to undertake the clean-up and remediation of contaminated sites in the region. Such an activity required engagement and coordination; however, public awareness about the deliverables of the project remained low. This lack of engagement has led to misunderstandings, whereby some stakeholders believe HYPREP will provide monetary compensation, not clean-up efforts. These types of misunderstanding can lead to mistrust and violence, yet can be mitigated through a structured stakeholder engagement process.

Conclusion

The issues of contaminated land in the Niger Delta are complex and the unintended consequences of contamination are often unimaginable. Local populations that are reliant on the land for agriculture or waterways for fishing are aware of the dangers that contamination presents, yet are unable to seek out alternative means to provide for themselves and their families. This can lead to frustration, and frustration can manifest in actions that lead to further contamination (e.g. sabotage). For policy to be effective, it must meet the needs of the population that it serves.

In Nigeria's quest to improve contaminated land management policies, it has chosen to adopt policies from the UK and the USA, where acute concern about the impacts of contaminated land no longer remains an issue. It is unreasonable to expect wholesale transfer of contaminated land policy to be effective. Based on the findings described in the case study, Nigeria would benefit from a form of contaminated land management triage that integrates elements of reporting, avoidance, remediation, and education in a framework that lends itself to the cultural uniqueness of the region. Delivery of such a policy will require a deeper understanding of the socio-cultural priorities and needs of affected stakeholders, and this can be delivered by a commitment to effective stakeholder engagement.

In this chapter, we have shown that stakeholder engagement is an established methodology that has been practised extensively by business to develop relationships based on the trust and transparency that allows businesses the licence to operate. More than a means to inform strategic development, stakeholder engagement has become a means for improving company valuations and is the norm for generating legitimacy with governments and communities. Governments too can benefit from stakeholder engagement, particularly in the development of policy where the motivation to balance budgets, build relationships, and generate legitimacy does not fundamentally differ from that of the firm. Although stakeholder engagement is not a panacea, we believe it can be an effective tool of the Nigerian Government to support policy development informed by societal needs in order to break the cycle of contamination that currently plagues the Niger Delta.

References

Ajayi, Dickson'Dare, and Chris O. Ikporukpo. 2005. An Analysis of Nigeria's Environmental Vision 2010. *Journal of Environmental Policy and Planning* 7 (4): 341–365.

Akpabio, E.M., A.S. Brown, I.E. Ansa, E.S. Udom, S. Abasi-ifreke, S. Eti-ido, ... and L.G.A. Ikono 2015. *Nigeria's Water and Sanitation: Spaces of Risk and the Challenges of Data.* Presented at the XVth World Water Congress, Vol. 25, p. 29, Edinburgh, Scotland.

Azapagic, Adisa. 2004. Developing a Framework for Sustainable Development Indicators for the Mining and Minerals Industry. *Journal of Cleaner Production* 12 (6): 639–662.

Bawole, Justice Nyigmah. 2013. Public Hearing or 'Hearing Public'? An Evaluation of the Participation of Local Stakeholders in Environmental Impact Assessment of Ghana's Jubilee Oil Fields. *Environmental Management* 52 (2): 385–397.

Benson, David, Irene Lorenzoni, and Hadrian Cook. 2016. Evaluating Social Learning in England Flood Risk Management: An 'Individual-Community Interaction' Perspective. *Environmental Science & Policy* 55: 326–334.

Bingham, Lisa Blomgren, Tina Nabatchi, and Rosemary O'Leary. 2005. The New Governance: Practices and Processes for Stakeholder and Citizen Participation in the Work of Government. *Public Administration Review* 65 (5): 547–558.

CERCLA, 2002. Comprehensive Environmental Response, Compensation and Liability Act of 1980.

Chinweze, Chizoba, Gwen Abiola-Oloke, and Chike Jideani. 2012. Oil and Gas Resources Management and Environmental Challenges in Nigeria. *Journal of Environmental Science and Engineering. B* 1 (4B): 535–542.

Cundy, A.B., R.P. Bardos, Andrew Church, M. Puschenreiter, W. Friesl-Hanl, I. Müller, S. Neu, M. Mench, Nele Witters, and Jaco Vangronsveld. 2013. Developing Principles of Sustainability and Stakeholder Engagement for "Gentle" Remediation Approaches: The European Context. *Journal of Environmental Management* 129: 283–291.

Devin, Bree L., and Anne B. Lane. 2014. Communicating Engagement in Corporate Social Responsibility: A Meta-level Construal of Engagement. *Journal of Public Relations Research* 26 (5): 436–454.

Eneh, Onyenekenwa Cyprian. 2011. Managing Nigeria's Environment: The Unresolved Issues. *Journal of Environmental Science and Technology* 4 (3): 250–263.

Etim, E.E., R. Odoh, A.U. Itodo, S.D. Umoh, and U. Lawal. 2013. Water Quality Index for the Assessment of Water Quality from Different Sources in the Niger Delta Region of Nigeria. *Frontiers in science* 3 (3): 89–95.

Fentiman, Alicia, and Nenibarini Zabbey. 2015. Environmental Degradation and Cultural Erosion in Ogoniland: A Case Study of the Oil Spills in Bodo. *The Extractive Industries and Society* 2 (4): 615–624.

FGN, 2004. National Water Policy Abuja, Nigeria. doi:https://doi.org/10.1093/chemse/bjt099.

Flak, Leif Skiftenes, and Jeremy Rose. 2005. Stakeholder Governance: Adapting Stakeholder Theory to E-government. *Communications of the Association for Information Systems* 16 (1): 31.

Forton, Osric Tening, Veronica E. Manga, Aaron S. Tening, and Akwinga V. Asaah. 2012. Land Contamination Risk Management in Cameroon: A Critical Review of the Existing Policy Framework. *Land Use Policy* 29 (4): 750–760.

Freeman, R. Edward. 2010. *Strategic Management: A Stakeholder Approach*. Minnesota: Cambridge University Press.

Garmendia, Eneko, and Sigrid Stagl. 2010. Public Participation for Sustainability and Social Learning: Concepts and Lessons from Three Case Studies in Europe. *Ecological Economics* 69 (8): 1712–1722.

Geaves, Linda H., and Edmund C. Penning-Rowsell. 2016. Flood Risk Management as a Public or a Private Good, and the Implications for Stakeholder Engagement. *Environmental Science & Policy* 55: 281–291.

Henisz, Witold J., Sinziana Dorobantu, and Lite J. Nartey. 2014. Spinning Gold: The Financial Returns to Stakeholder Engagement. *Strategic Management Journal* 35 (12): 1727–1748.

Idemudia, Uwafiokun. 2014. Corporate-community Engagement Strategies in the Niger Delta: Some Critical Reflections. *The Extractive Industries and Society* 1 (2): 154–162.

Ite, Aniefiok E., Udo J. Ibok, Margaret U. Ite, and Sunday W. Petters. 2013. Petroleum Exploration and Production: Past and Present Environmental Issues in the Nigeria's Niger Delta. *American Journal of Environmental Protection* 1 (4): 78–90.

Jenkins, Heledd, and Natalia Yakovleva. 2006. Corporate Social Responsibility in the Mining Industry: Exploring Trends in Social and Environmental Disclosure. *Journal of Cleaner Production* 14 (3–4): 271–284.

Jha, Anand, and James Cox. 2015. Corporate Social Responsibility and Social Capital. *Journal of Banking & Finance* 60: 252–270.

Kadafa, Adati Ayuba. 2012. Oil Exploration and Spillage in the Niger Delta of Nigeria. *Civil and Environmental Research* 2 (3): 38–51.

Krippendorff, Klaus. 2012. *Content Analysis: An Introduction to Its Methodology*. Thousand Oaks: Sage.

Moomen, Abdul–Wadood, and Ashraf Dewan. 2017. Probing the Perspectives of Stakeholder Engagement and Resistance Against Large-Scale Surface Mining in Developing Countries. *Corporate Social Responsibility and Environmental Management* 24 (2): 85–95.

Mzembe, Andrew Ngawenja. 2016. Doing Stakeholder Engagement Their Own Way: Experience from the Malawian Mining Industry. *Corporate Social Responsibility and Environmental Management* 23 (1): 1–14.

NDDC, 2006. *Niger Delta Development Master Plan 2006.* http://www.nddc. gov.ng/NDRMPChapter 1.pdf. Accessed 1 Sep 2016.

Noy, Chaim. 2008. Sampling Knowledge: The Hermeneutics of Snowball Sampling in Qualitative Research. *International Journal of Social Research Methodology* 11 (4): 327–344.

Nriagu, Jerome, Emilia A. Udofia, Ibanga Ekong, and Godwin Ebuk. 2016. Health Risks Associated with Oil Pollution in the Niger Delta, Nigeria. *International Journal of Environmental Research and Public Health* 13 (3): 346.

Nwankwoala, H.O. 2014. Problems and Options of Integrated Water Resources Management in Nigeria: Administrative Constraints and Policy Strategies. *International Letters of Natural Sciences* 9: 12–25.

Nwilo, P.C., and O.T. Badejo. 2005. Oil Spill Problems and Management in the Niger Delta. In *International Oil Spill Conference* 2005 (1): 567–570. Washington, DC: American Petroleum Institute.

Nwilo, Peter C., and Olusegun T. Badejo. 2006. Impacts and Management of Oil Spill Pollution Along the Nigerian Coastal Areas. *Administering Marine Spaces: International Issues* 119: 1–15.

Okeke, P.N., and C.Q. Okpala. 2014. Effects of Gas Flaring on Selected Arable Soil Quality Indicators in the Niger Delta, Nigeria. *Sky Journal of Soil Science and Environmental Management* 3 (1): 001–005.

OPEC. 2015. *Nigeria: Facts and Figures*. http://www.opec.org/opec_web/en/about_us/167.htm. Accessed 1 Sep 2016.

Oyebamiji, M. Adekola, and C. Igwe Mba. 2013. Effects of Oil Spillage on Community Development in the Niger Delta Region: Implications for the Eradication of Poverty and Hunger (Millennium Development Goal One) in Nigeria. *World Journal of Social Science* 1 (1): 27.

Pegg, Scott, and Nenibarini Zabbey. 2013. Oil and Water: The Bodo Spills and the Destruction of Traditional Livelihood Structures in the Niger Delta. *Community Development Journal* 48 (3): 391–405.

Péry, A.R.R., G. Schüürmann, Philippe Ciffroy, Michael Faust, T. Backhaus, Lothar Aicher, Enrico Mombelli, et al. 2013. Perspectives for Integrating Human and Environmental Risk Assessment and Synergies with Socio-economic Analysis. *Science of the Total Environment* 456: 307–316.

Prpich, George, Jens Evans, Phil Irving, Jerome Dagonneau, James Hutchinson, Sophie Rocks, Edgar Black, and Simon J.T. Pollard. 2011. Character of Environmental Harms: Overcoming Implementation Challenges with Policy Makers and Regulators. *Environmental Science & Technology* 45 (23): 9857–9865.

Ramirez-Andreotta, Monica D., Mark L. Brusseau, Janick F. Artiola, Raina M. Maier, and A. Jay Gandolfi. 2014. Environmental Research Translation: Enhancing Interactions with Communities at Contaminated Sites. *Science of the Total Environment* 497: 651–664.

Rangarajan, Kiran, Suzanna Long, Alan Tobias, and Marie Keister. 2013. The Role of Stakeholder Engagement in the Development of Sustainable Rail Infrastructure Systems. *Research in Transportation Business & Management* 7: 106–113.

Reed, Mark S. 2008. Stakeholder Participation for Environmental Management: A Literature Review. *Biological Conservation* 141 (10): 2417–2431.

Rizzo, Erika, Marco Pesce, Lisa Pizzol, Filip Mihai Alexandrescu, Elisa Giubilato, Andrea Critto, Antonio Marcomini, and Stephan Bartke. 2015. Brownfield Regeneration in Europe: Identifying Stakeholder Perceptions, Concerns, Attitudes and Information Needs. *Land Use Policy* 48: 437–453.

Rowe, Gene, and Lynn J. Frewer. 2005. A Typology of Public Engagement Mechanisms. *Science, Technology, & Human Values* 30 (2): 251–290.

Sam, Kabari, Frédéric Coulon, and George Prpich. 2017a. Use of Stakeholder Engagement to Support Policy Transfer: A Case of Contaminated Land Management in Nigeria. *Environmental Development* 24: 50–62.

————. 2017b. Management of Petroleum Hydrocarbon Contaminated Sites in Nigeria: Current Challenges and Future Direction. *Land Use Policy* 64: 133–144.

————. 2017c. A Multi-attribute Methodology for the Prioritisation of Oil Contaminated Sites in the Niger Delta. *Science of the Total Environment* 579: 1323–1332.

Sardinha, Idalina Dias, Daniela Craveiro, and Sérgio Milheiras. 2013. A Sustainability Framework for Redevelopment of Rural Brownfields: Stakeholder Participation at SÃO DOMINGOS Mine, Portugal. *Journal of Cleaner Production* 57: 200–208.

Scholl, Hans J. 2004. Involving Salient Stakeholders: Beyond the Technocratic View on Change. *Action Research* 2 (3): 277–304.

Slinger, Giles. 1999. Spanning the Gap—The Theoretical Principles That Connect Stakeholder Policies to Business Performance. *Corporate Governance: An International Review* 7 (2): 136–151.

Snape, Dee, Jamie Kirkham, Jenny Preston, Jennie Popay, Nicky Britten, Michelle Collins, Katherine Froggatt, et al. 2014. Exploring Areas of Consensus and Conflict Around Values Underpinning Public Involvement in Health and Social Care Research: A Modified Delphi Study. *BMJ Open* 4 (1): e004217.

UNEP. 2011. *Environmental Assessment of Ogoniland*. UNEP, Switzerland.

Vogel, David. 2007. *The Market for Virtue: The Potential and Limits of Corporate Social Responsibility*. Washington, DC: Brookings Institution Press.

Wang, Zhihong, and Joseph Sarkis. 2017. Corporate Social Responsibility Governance, Outcomes, and Financial Performance. *Journal of Cleaner Production* 162: 1607–1616.

Wheeler, David, Heike Fabig, and Richard Boele. 2002. Paradoxes and Dilemmas for Stakeholder Responsive Firms in the Extractive Sector: Lessons from the Case of Shell and the Ogoni. *Journal of Business Ethics* 39 (3): 297–318.

World Bank. 2016. World Development Indicators. http://data.worldbank.org/data-catalog/world-development-indicators. Accessed 1 Nov 17.

Zhao, Dong, Andrew P. McCoy, Brian M. Kleiner, Thomas H. Mills, and Helen Lingard. 2016. Stakeholder Perceptions of Risk in Construction. *Safety Science* 82: 111–119.

5

Cooperatives' Potential to Diffuse Appropriate Solar Technologies in Uganda

Ahmed Kitunzi Mutunzi and Shailendra Vyakarnam

Introduction

Although there is plentiful sunshine across much of Africa, less than 25% of the population effectively utilize solar energy to address their need for electricity (International Energy Agency 2015). Solar energy boosts countries' energy security through reliance on an inexhaustible and import-independent resource. It also increases sustainability, lowers the costs of mitigating climate change, reduces pollution, and keeps fossil fuel prices lower than otherwise might be the case (Ondraczek 2013). Cutting-edge solar technologies are progressively inexpensive, accessible, and manageable for alleviating numerous socio-economic and environmental challenges (Bradford 2014).

A. Kitunzi Mutunzi (✉)
Makerere University, Kamapala, Uganda
e-mail: amutunzi@bams.muk.ac.ug

S. Vyakarnam
Cranfield University, Bedford, UK
e-mail: shailendra.vyakarnam@cranfield.ac.uk

© The Author(s) 2019
S. Adesola, F. Brennan (eds.), *Energy in Africa*,
https://doi.org/10.1007/978-3-319-91301-8_5

The scarcity of electricity causes excessive reliance on fossil fuel for lighting; such fuel is often expensively imported and has hazardous effects on people's health through inhaled fumes (Eder et al. 2015). These circumstances aggravate global warming through increased emission of carbon dioxide, intense poverty, unemployment, and limiting productivity (IEA 2011). Lack of electricity also impedes social interaction and educational uptake (Barnes 2007).

In Uganda, over 85% of the population do not access grid power and thus depend on non-renewable sources of energy, particularly wood, and fossil fuels. Such prevalent sources of energy are gravely hazardous to the physical environment, economy, and people's health and productivity (Da Silva et al. 2014). Furthermore, only 15% of the population, mainly in urban areas, access the grid while facing irregular and unreliable electricity supplies. Grid-based hydroelectric power costs 18 US cents per unit, which makes it unaffordable for over 95% of the communities, especially the rural poor (Bizzari 2009).

Many people often spend much of their productive time and financial resources on collecting firewood and kerosene for cooking and lighting in Uganda, Ghana, and other countries of Sub-Saharan Africa (Piggins 2014). Also as Uganda's population increases the demand for wood fuel escalates, and this leads to extensive deforestation that may eventually lead to desertification (Twaha et al. 2012). Kerosene lanterns are among the commonest type of lamps used by people who do not access grid electricity. These lamps are relatively dim and emit unhealthy fumes (Neelsen and Peters 2011). Addressing such social and environmental challenges efficiently and sustainably is increasingly handled by social enterprises, such as Cooperatives (Wimmer 2012).

Over 70% of the rural communities are sparsely settled and over 20% of these are nomadic or semi-nomadic pastoralists in many countries of Sub-Saharan Africa (Mann et al. 2014). Consequently, the most appropriate means of alleviating energy crises in Uganda is ostensibly the diffusion of low-cost, robust off-grid solar technologies, which thus forms the focus of this chapter.

The Context of the Study

Appropriate solar technologies can alleviate energy-related challenges in Africa as the continent receives abundant and intense sunshine on most days of the year (Bradford 2014). There is a growing industry engaged in manufacturing and distribution of increasingly affordable, robust, and user-friendly solar technologies that are seemingly appropriate for Africa. There are solar products for lighting, entertainment, and phone charging at approximate US $25 per unit. Many companies have been attempting to sell solar technology in Africa for more than five years now, but with limited impact (IEA 2011).

While most solar equipment vendors in Africa are profit oriented, their products are socially and environmentally essential in Africa (Karekezi et al. 2003). Their business model is generally to take substantial upfront deposit payments for their products with follow-on credit terms met via mobile phone payments. However, the targeted customers for solar equipment largely lack the financial resources and reliable mobile phone networks that are used to buy the vitally needed, and often comparatively expensive, solar equipment being offered, although as we have seen, the costs are tumbling (Mawejje and Okumu 2014). These factors, among others, are apparently responsible for the low diffusion of solar technologies.

It is therefore vital to consider disseminating modern solar technologies through a different business model that is more socially oriented and capable of financially facilitating its customers to buy what they need. This chapter thus delves into the potential of Savings and Credit Cooperatives (SACCOs) to play this role.

A SACCO is a member-driven, democratic, unique, self-help, mutual initiative that is owned, managed, and governed by its members who have a mutual bond such as working for the same employer, belonging to the same social fraternity, labour union, church or mosque, or living in the same community. Membership is open to all who belong to the group, the primary purpose being to save their money together in the SACCO and to extend loans to each other at fair interest rates. Interest charged on loans is usually used to cover administration costs and there is no payment or profit extended to outside interests or internal owners. The members are the owners, and it is they who decide on the allocation of their money (SACCOL 2015). These basic virtues of SACCOs seemingly qualify them to serve as suitable conduits for spreading socio-economically vital items such as innovative solar technologies.

In Uganda, Cooperatives have been very influential and instrumental in the socio-economic development of the country for over 100 years (Kyazze 2010). By 1900, Cooperatives were functional in various parts of Uganda as informal but well-organized and socio-economically developmental establishments. Cooperatives in Uganda acquired formal status with the advent of the Cooperative Ordinance of 1946 and the Cooperative Societies Act of 1962. Between 1962 and 1977, the performance of Cooperatives in Uganda was particularly remarkable, with the Government offering them a monopoly status in agricultural marketing. Between 1977 and 1999, however, the performance and significance of Cooperatives in Uganda were severely shaken by political meddling, corruption, indebtedness, and grave mismanagement. These devastations led to the disbandment of many Cooperative societies in the country (Kyazze 2010). Since 2006, the Cooperative movement in Uganda has experienced a revival with considerable support and facilitation from the Government and development partners.

Today Uganda boasts of having over 13,179 formally registered Cooperatives, and over 60% of these are SACCOs (Ahimbisibwe 2013). There has been a formation of specialized innovative cooperative enterprises such as rural water, rural electrification, and indigenous community service farms that can boost the socio-economic development of Uganda (Kyazze 2010). The most dominant and influential forms of specialized Cooperatives in Uganda are SACCOs, as they constitute over 60% of the formally recognized Cooperatives (Ahimbisibwe 2013). SACCOs are conventional Cooperatives established to the principles, values, and ethical guidelines decreed by the International Cooperative Alliance (ICA). SACCOs are channels for facilitating citizens financially, to enable them to address their various socio-economic challenges (Mpiira et al. 2014). Hence, there is a countrywide propagation of SACCOs in Uganda, organized under an umbrella organization known as the Uganda Cooperative Savings and Credit Union (UCSCU), which serves as a Governmental regulatory agency. These facts motivated us to explore the feasibility of SACCOs as channels of distribution for appropriate solar technologies.

Methodology

This study is principally a cross-sectional survey that predominantly utilizes a triangulation of qualitative techniques for collecting and analysing data as diffusion of innovations is hard to quantify due to the complexity of humans and human networks. Primary data for the study were gathered through focus group discussions, a few interviews, minimal observations, and a review of the relevant literature. The key informants for both the interviews and focus group discussions were purposively selected from various categories of stakeholders of the solar industry, such as the prevailing solar vendors and merchants, policymakers, solar energy users, and regulatory authorities. Observations focusing on the selling and uptake of solar technologies were also selectively made as part of the field surveys carried out in both Uganda and Ghana. The major sources of secondary data reviewed for this chapter are the existing literature on the theory of diffusion of innovations (Rogers 2003; Miller 2012); the attributes of savings and credit Cooperatives (Ahimbisibwe 2013; Birchall 2004; Kyazze 2010; MacPherson 1995); and the incumbent business models for selling solar technology in Uganda (Mawejje and Okumu 2014; Piggins 2014). The study entailed surveys conducted in Ghana so as to augment the chapter with a comparative analysis of the solar industry in at least two typical African countries. Data collected were analysed qualitatively using thematic, deductive, and inductive approaches.

We compared our findings with the literature on solar technology adoption in Africa, and by considering the issues raised in the adoption of innovation, we have brought together a simple menu of findings and recommendations which are articulated below. In our recommendations for scaling up efforts, we locate the findings in the context of the innovation journey put forward in Phadke and Vyakarnam (2017).

Objectives of the Data Collection

"Potential of the Cooperatives (SACCOs) to sell and diffuse appropriate solar technologies in Uganda" was the main objective of the survey that, thus, guided our focus group discussions and interviews. We set out our findings from the key informants (i.e. focus groups and interviewees) below.

Defining Appropriate Solar Technologies for Uganda

Before we entered the substance of the focus groups and interviews, we asked our respondents to help define what they believed was appropriate solar technology.

Results from most of the respondents with whom we engaged indicate that appropriate solar technologies are defined as solar-powered equipment that are affordable, robust, resilient for rough conditions, simple, and user-friendly for users to install, operate, and maintain with little or no need for hired technicians. Such equipment also ought to be of the highest level of utility and be able to solve day-to-day problems, such as lighting and phone charging, and increasingly able to power televisions.

We found that our respondents in Uganda and Ghana had similar requirements. Having defined the "product" attributes, we progressed to explore what it might take to encourage the wider adoption of solar technologies.

Prerequisites for the Diffusion of Appropriate Solar Technologies in Uganda

Our initial perspective was that solar technologies were rather slow to be accepted in Uganda. And this is broadly in keeping with technology adoption theories (Rogers 2003), which propose that for a technology to diffuse, the market needs to adopt it and that there are three categories of customers: the so-called innovators/early adopters who have the resources and capability to try new technology and do not need to be fully convinced before they try; then comes another category of customers dubbed the early majority—this category helps increase uptake as they become increasingly convinced by the efficacy and, perhaps, as prices begin to drop; and finally for adoption, there is the late majority as the product and market begin to mature (Rogers 2003).

We find that in Uganda, the innovators and early adopters are mainly in the cities where electricity is erratic rather than unavailable. Our focus is on the vast majority that is, in rural and peri-urban Uganda, a low-income population. The findings, based somewhat on the notion of crossing chasms (Phadke and Vyakarnam 2017), indicate that we should ask the question about what is needed to shift our products and services from the early adopters to the majority of customers. There is a clear inflection point—when sales stall because we are unable to convince the majority to buy (Phadke and Vyakarnam 2017). The assumption is that there are barriers to majority adoption and that is what we explore in our study.

Mass Sensitization and Awareness According to Miller (2012), multitudes of people, especially the rural poor, need to be well sensitized, educated, informed, and/or made aware of the existence, accessibility, utilization (i.e. application/operation), and maintenance of appropriate solar technologies for their energy needs. This was generally noted as a basic requirement for the countrywide demand for and uptake of modern solar technologies. In our conversations, we found that there was a misunderstanding about the technologies on the one hand and market entry of low-quality technology that could not sustain long-term usage.

Financial Facilitation According to Miller (2012), affordable financing and credit schemes need to be established to enable all potential and prospective customers, especially the poor rural communities, to buy appropriate solar technologies. Miller (2012) defines this as long-term soft loans, grants from Governments and development agencies, and fair credit terms extended by the vendors of the solar equipment.

Conducive Laws, Policies, and Regulations Specific policy recommendations made by vendors, agencies, and users in our focus groups were to eliminate all forms of taxation on solar technologies, and to have regulations to determine standards for solar technologies being imported into the country to ensure customers' protection. These findings are also corroborated by Miller (2012) who remarks that Government policies that espouse the utilization of solar technologies are crucial for diffusing solar technologies into the poor rural communities in emerging markets.

Repositioning of Solar Technologies Some of the key informants we consulted observed that there is a need for all key stakeholders of the solar industry to regard and treat solar technologies as social and/or public goods, rather than commercial items. This entails Government intervention by exempting taxation of any solar equipment and expanding its distribution channels so as to make it more affordable and accessible countrywide. They note that such measures may drastically reduce the cost of solar technology, make it more accessible, and thus boost its diffusion across the whole country.

Quality Defined as Simple and Affordable Solar Technologies Many respondents asserted that there was a "swamping" of the country with poor quality, overly complicated, expensive, at times delicate, unreliable, unsustainable, and thus very disappointing range of products. This, perhaps, explains the current modest uptake of solar technologies in Uganda. It is important therefore to have Government-driven standards for good quality (i.e. fit for the purpose), to achieve the diffusion of solar technology into the country.

Credible and Accessible Vendors/Suppliers of Solar Technologies Our respondents expressed strong views about the characteristics they wanted from their solar vendors. They need vendors who are technically proficient, reachable, knowledgeable, approachable, honest, plausible, convincing, responsive, trustworthy, reliable, and believable. There is also a need for warranties and guarantees for the solar technologies. Many respondents claimed that the prevailing low uptake of solar technologies in Uganda may be due to vendors not being well-informed and being located mainly in urban or semi-urban centres, while the majority of potential customers live in rural areas. They felt the vendors were more focused on profit-maximization rather than on customer satisfaction.

Widespread Technical Maintenance/Service Centres Several key informants disclosed that there is need to establish accessible countrywide technical support for the installation, servicing or maintenance, and sustainability of solar technologies. From the surveys conducted, it was noted that the present poor uptake of solar technology is partly attributed to failure to access technical assistance, especially for the rural-based customers.

The factors explained above are similar to those revealed by the comparative studies conducted in Ghana and the literature reviewed. We now look at our findings regarding the suitability of SACCOs to act as channels for the diffusion of solar-powered technologies.

The Strengths of Cooperatives in Diffusing Appropriate Solar Technologies in Uganda

Findings from the empirical surveys conducted in Uganda indicate that SACCOs have the following strengths and/or relative advantages in acting as channels for selling and diffusing solar-powered technologies:

Community Outreach Respondents generally observed that, compared to the current marketing channels for solar technologies, SACCOs have a much broader outreach as they are increasingly established and scattered countrywide, networked, and have large memberships of over 250 households on average per SACCO.

Financial Facilitation SACCOs provide credit facilities at less than 15% per annum compared to banks that operate at a minimum of 35% per annum (Munyambonera and Adong 2013). Respondents divulged that beyond banks, there are private sector lenders who charge even higher rates. SACCOs mitigate the lending risks by virtue of being peer lending agencies and can, therefore, afford to charge much less and in keeping with their mandate with members. Private vendors operate on the basis of an upfront payment and regular mobile payments to link with micro-credit formulas, where affordable cash flow is taken into account rather than the interest that is charged. One of the incumbent solar vendors we met requires their customers to make an upfront payment of not less than 50% for whatever solar equipment they wish to purchase. The balance of the cost is then paid via mobile phone networks in equal instalments spread over a period of not more than two years. An unintended consequence has been that customers borrow money for the upfront payment and the project then becomes a serious burden.

Communication Channels Respondents largely state that SACCOs efficiently enable the flow of information amongst all key stakeholders of the solar industry. This strength arises from the fact that SACCOs are widely networked and their respective members often meet and share critical information regarding their Cooperative's business. Hence, SACCOs can utilize this comparative advantage to easily create awareness, sensitization, and marketing of solar technologies.

Credibility and Integrity Respondents also indicated that SACCOs are generally well trusted, respected, and seen to be reliable by their members. They are seen as social enterprises with the improvement of the socio-economic welfare of their members as their core purpose, rather than making profits at their expense.

Credit Rating of Prospective Customers The key informants we engaged generally opined that SACCOs can more easily assess and determine the credit rating of their respective members, which should give confidence to vendors that they will be repaid for their products and services.

The strengths of SACCOs derived from empirical surveys in Uganda, as described above, are buttressed by findings from the literature reviewed in respect to the suitability of SACCOs to diffuse solar technologies, as explained below. According to the literature reviewed, the comparative advantages of utilizing SACCOs to diffuse solar technologies can be appreciated from the perspective of their universal characteristics as well as a Uganda-specific standpoint as discussed below.

In Uganda, the Government, development agencies, and civil society increasingly support, facilitate, bolster, uphold, and endorse SACCOs as exceptional developmental vehicles (Mpiira et al. 2014). The SACCOs have progressively served as prime organizations for the financial facilitation of Ugandans, especially the poor rural communities which constitute the majority of ordinary people, to meet their various socio-economic needs (Ahimbisibwe 2013). As such, SACCOs in Uganda have numerous and distinctive advantages that make them viable channels for diffusing appropriate solar technologies to the communities as reviewed later.

Systematic Propagation The Government and development partners in Uganda are increasingly formulating considered policies and schemes for establishing SACCOs in all regions and socio-economic sectors of Uganda. The process of registering SACCOs is very simple as Cooperatives in Uganda have a specific office for expediting the formal registration of all Cooperatives (Ahimbisibwe 2013). In 2008, the Ugandan Government decided to facilitate the establishment of SACCOs in all parts and sectors of the country (Ahimbisibwe 2013). By 30 August 2012, Uganda had more than 5220 SACCOs registered as formal entities (Ahimbisibwe 2013). Thus this deliberate establishment of SACCOs all over the country makes these Cooperatives suitable structures for diffusing appropriate solar technologies in Uganda.

Ample Financial and Technical Assistance The Ugandan Government and its development partners do progressively provide substantial technical and financial reinforcement to the SACCOs in Uganda (Beisland and Mersland 2012); for example, they have mobilized over US $500 million for the micro-financing of Ugandans directly and indirectly through SACCOs (Munyambonera and Adong 2013). Development agencies that have funded the technical and financial capacities of SACCOs in Uganda include the Swedish International Development Cooperation Agency, the World Council of Credit Unions, and the United Kingdom Department for International Development. The Ugandan Government disburses financial assistance through SACCOs to its citizens, especially the poor rural communities (Ahimbisibwe 2013). Thus, SACCO members are among the few people in Uganda who can access soft and collateral-free loans for meeting their socio-economic needs, such as clean energy (Munyambonera and Adong 2013).

Collateral: Free and Inexpensive Financial Facilities The financial services offered by SACCOs in Uganda are the cheapest and easiest to access in the country as their members do not have to borrow against any tangible collateral other than their social networks and credibility. The SACCOs provide financial facilitation to their members at concessional rates, unlike most commercial banks and money lenders in Uganda (Ahimbisibwe 2013). Thus, many people are motivated to join SACCOs as loyal members and this enhances the potential of SACCOs to diffuse solar technologies in Uganda.

Favourable Governmental Support and Enablement The Ugandan Government directly supports the establishment, operations, and sustenance of SACCOs by constituting several organizations to monitor, regulate, finance, oversee, or manage the establishment of well-managed SACCOs in Uganda. Agencies formed to monitor, regulate, support, and facilitate SACCOs include the UCSCU, Uganda Microfinance Support Centre, and Uganda Cooperatives Alliance, among others (Munyambonera and Adong 2013).

Noble Corporate Values, Ethics, and Principles (Good Corporate Governance Grounds) The International Cooperatives Alliance (ICA) instituted very impressive moral values, standards, and principles for all Cooperatives, including SACCOs (Phelan et al. 2012). If SACCOs uphold these exceptional values, they enhance the SACCOs' viability to serve as conduits for solar technologies.

Notwithstanding the numerous strengths of Cooperatives analysed above, SACCOs have some distinct weaknesses and limitations. They also face challenges which might impede them as organizations for the diffusion of solar technologies. Results from the empirical surveys point to the following issues.

Focus on Savings and Credit Services Respondents generally noted that, currently, SACCOs concentrate on their core business of managing financial savings and credit services for their respective members. Hence, a diversification into the provision of solar technologies may well place them at risk regarding their core activity.

Technical Proficiencies According to most of the respondents contacted, SACCOs lack human resources with the technical and selling skills, competencies, and knowledge that are necessary for the diffusion of appropriate solar technologies.

Basic Facilities/Resources Some respondents remarked that SACCOs lack the necessary logistics and secure storage spaces to hold inventory. SACCOs also lack the ability to handle returns of any faulty equipment.

Mandate The management of SACCOs with whom we interacted stated that they were not mandated to sell or deal with solar technologies, although existing laws and regulations implicitly allow for such transactions. Therefore, there is either a need for a specific and explicit Act or the clarification of existing laws to empower management to take action.

Members' Behaviours According to some of the respondents, some SACCO members lack honesty, integrity, responsiveness, and commitment to all the causes of their respective Cooperatives. Thus, the fear is the risk of default on loans or that they take advantage of affordable technologies and turn it into profiteering ventures. However, some respondents observed that ethical behaviour among SACCO members is generally being increasingly enforced by the UCSCU as one of its mandates.

The literature highlighted the following weaknesses of SACCOs:

Business Acumen SACCOs do not have good mercantile/commercial perspicacity and acumen, such as keeping business secrets, or forceful marketing strategies that provide competitive and survival edges for most business entities (Ahimbisibwe 2013). This may be due to the supreme principles and ethical values imposed by the ICA, which emphasizes utmost transparency, focus on social motives and free access to information by all stakeholders. Lack of business acumen combined with ICA values may compromise the capacity of SACCOs to engage in diffusing solar technologies.

Organizational Management, Leadership, and Governance Most SACCOs are managed by their respective members, the majority of whom lack professional management skills (Mpiira et al. 2014). Such poor management leads to frequent fraud, dysfunctional conflicts, and dishonesty, among other unethical practices (Mpiira et al. 2014). Nevertheless, one of the key roles of the UCSCU is to improve the management, leadership, and governance of all SACCOs through training and other programmes (Ahimbisibwe 2013).

ICA Regulations and Principles A major cause of weaknesses and drawbacks for SACCOs is their apparent failure to adhere to the universal principles, ethics, values, declarations, and standards for all Cooperatives, including SACCOs, that have been established by the ICA. Some of the notable contraventions of ICA recommendations include the violation of Cooperatives' values of honesty and openness (Munyambonera and Adong 2013).

In summary, the literature indicates the outstanding weaknesses of all Cooperatives, as being their vulnerability to weak corporate governance, poor leadership, and poor economic activity by their members. SACCOs particularly suffer from political interference, lack of cooperative member education, and a widespread culture of poor savings. Other challenges include massive fraud, high levels of delinquency, and inadequate security for guarding SACCO premises (Ahimbisibwe 2013). These documented challenges pose a considerable threat to the prosperity of SACCOs and other types of Cooperatives. Hence, these identified weaknesses inhibit the potential of SACCOs to sell and diffuse appropriate solar technologies in Uganda. However, several interventions have been arranged by the Government and development agencies to mitigate these challenges (Ahimbisibwe 2013) and we can look to a more optimistic future.

Conclusions

At the Micro Level Our Findings Suggest

A product range that meets daily requirements and is therefore fit for purpose comprises solar equipment that is simple to access, install, and operate. There is a need for reliable technical maintenance and servicing. Products should go through an approval process so that customers are reassured that they have genuine products. And finally, they need financial assistance to acquire the technologies.

At the Meso Level—Where SACCOs Can Play a Role—Our Findings Suggest

The above criteria are core requirements of users. In addition to these, at a meso level what is needed appears to be a much greater awareness-raising campaign that promotes solar technologies. It is in this context that SACCOs provide a potential solution as they are a Government-approved agency. If they can be empowered to diffuse solar technology solutions for their members, the whole sector can benefit from their wide reach, ability to provide financial aid to their members, and draw on Government support, as well as fit into the spirit of solar technologies as a social good. It is this agency role, we would argue, which has the capacity to scale up the adoption of solar technologies.

SACCOs do of course face challenges as we have observed above. They are "bankers" rather than "retailers" and as a result, they do not have the skills, knowledge, resources, physical attributes, and business acumen to build this activity. They are vulnerable to poor governance procedures and unreliable members who may default on their loans, although this weakness is being ably addressed by the UCSCU. But these are all issues that, with training and investment, can be solved. It is also possible to develop a business model that brings SACCOs together with reliable private sector players who have the knowledge and know-how for the deployment of solar technologies.

At a Macro Level

Appropriate solar technologies ought to be identified, branded, and positioned as essential social goods that serve as solutions to the socio-economic and environmental challenges in Uganda and the rest of Africa. This calls for the repositioning of solar innovations as tax-free, low-cost, social items, rather than conventional commercial (for-profit) merchandise. Hence, practical Governmental policies, laws, and regulations, plus interventions by development partners, are necessary for rebranding, availing and popularizing solar technologies as social goods. Recommendable regulations and policies here include establishing strict quality assurance authorities to ensure importation, sale, and distribution of only high quality, branded solar technologies from reputable, responsible, responsive, and accountable sources or suppliers. These measures are bound to facilitate making appropriate solar technologies more affordable and readily accessible by all those who need them, thereby increasing their diffusion in Uganda and the rest of Africa. Policies may also be considered that encourage local businesses to invest in partial assembly or other value-adding roles that can also bring down the costs of imports.

In addition to these raised costs are the consequences of import bills, as the Uganda shilling continues to depreciate.

Recommendations for Further Studies

There are significant numbers of issues and stakeholders that need to be investigated. Vendors, manufacturers, and others in the value chain need to be better understood regarding their motivations, barriers, and enablers. Development agencies can contribute to macro-policy recommendations to scale up the adoption of solar technologies. And a wider study of other regions is needed to understand policies, fiscal and other arrangements that have been adopted in order to replicate the best lessons. Such studies are likely to generate lessons that may enlarge the knowledge provided by this chapter.

Acknowledgements We gratefully acknowledge the sponsorship of ALBORADA and Cambridge-Africa Program for Research Excellence (CAPREX) which has enabled the development of this chapter. We are also grateful to all key informants or respondents in Ghana and Uganda.

References

Ahimbisibwe, Fred. 2013. *Overview of Cooperatives in Uganda. Ministry of Trade, Industry, and Cooperatives.* Retrieved from http://www.mtic.go.ug/index.php?/cat_view/129-trade/Page-4/.

Barnes, Douglas F., ed. 2007. *The Challenge of Rural Electrification: Strategies for the Developing Countries*. Earthscan. Retrieved from http://www.amazon.com/The-Challenge-Rural-Electrification-Strategies/dp/19331154.

Beisland, L.A., and R. Mersland. 2012. The Use of Microfinance Services Among the Economically Active Disabled People: Evidence from Uganda. *Journal of International Development* 24 (S1): S69–S83. Retrieved from http://onlinelibrary.wiley.com.

Birchall, Johnston. 2004. *Cooperatives and the Millennium Development Goals*. Geneva: International Labour Organization.

Bizzari, M. 2009. Safe Access to Firewood and Alternative Energies in Uganda: An Appraisal Report. *WFP*, Rome. Retrieved from http://www.genderconsult.org/res/doc/SAFE_Uganda.pdf. Retrieved from http://onlinelibrary.wiley.com/doi/10.1002/jid.1720/abstract.

Bradford, Travis. 2014. *Solar Energy: A BIT of Solar Revolution*. Cambridge: MIT Press.

Da Silva, I.P., G. Batte, J. Ondraczek, G. Ronoh, and C.A. Ouma. 2014. Diffusion of Solar Energy Technologies in the Rural Africa: Trends in Kenya and the LUAV Experience in Uganda. *1st Africa Photovoltaic Solar Energy Conference & the Exhibition Proceedings* 1 (1): 106, May. Retrieved from http://www.researchgate.net/publication/263078787_Publication_Diffusion_of_solar.

Eder, J.M., C.F. Mutsaerts, and P. Sriwannawit. 2015. Mini-grids and Renewable Energy in Rural Africa: How Diffusion Theory Explains the Adoption of Electricity in Uganda. *Energy Research & Social Science* 5: 45–54. Retrieved from https://doi.org/10.1016/j.erss.2014.12.014.

IEA. 2011. *The "World Energy Outlook 2011"*. Retrieved June 24, 2014 from http://lightsforbillions.com/category/information/.

International Energy Agency. 2015. *World Energy Outlook-2014*. Retrieved from http://www.iea.org/bookshop/477-World_Energy_Outlook_2014.

Karekezi, S., W. Kithyoma, and Energy Initiative. 2003. Renewable Energy Development. In a *Workshop on African Energy Experts on Operationalizing the NEPAD Energy Initiative*, June, pp. 2–4. Retrieved from http://www.gubaswaziland.org/files/documents/resource10.pdf.

Kyazze, Lawrence M. 2010. Cooperatives: The Sleeping Economic and Social Giants in Uganda – *Series on the Status of Cooperative Development in Africa*. www.ilo.org/coopafrica.

MacPherson, Ian. 1995. *Co-operative Principles for the 21st Century*. Vol. 26. Geneva, International Co-operative Alliance.

Mann, David., Josephine Namukisa, and Alex Ndibwami. 2014. Energy and Urbanization in Uganda: Context Report and Literature Review. Uganda Martyrs University—Faculty of the Built Environment. Retrieved from http://samsetproject.net/wp-content/uploads/2016/02/SAMSET_Lit-Rev_ Uganda_FINAL_25-April-2014.pdf.

Mawejje, Joseph, and Ibrahim M. Okumu. 2014. *Tax Evasion and the Business Environment of Uganda*. Retrieved from http://makir.mak.ac.ug/handle/10570/4047.

Miller, D. 2012. *Selling Solar: Diffusion of Renewable Energy in the Emerging Markets*. London: Routledge.

Mpiira, S., B. Kiiza, E. Katungi, J.R.S. Tabuti, C. Staver, and W.K. Tushemereirwe. 2014. Determinants of Net Savings Deposits held in the Savings and Credit Cooperatives (SACCO's) in Uganda. *Journal of Economics and International Finance* 6 (4): 69–79.

Munyambonera, Ezra, and Annet Adong. 2013. *Access and the Use of Credit in Uganda: Unlocking a Dilemma of Financing Small Holder Farmers*. Retrieved from http://onlinelibrary.wiley.com/book/10.1002/9780470674871.

Neelsen, S., and J. Peters. 2011. Electricity Usage in Micro-Enterprises—Evidence from Lake Victoria, Uganda. *Energy for Sustainable Development* 15 (1): 21–31.

Ondraczek, J. 2013. The Sun Rises in the East (of Africa): A Comparison of the Development and Status of Solar Energy Markets in Kenya and Tanzania. *Energy Policy* 56: 407–417.

Phadke, Uday, and Shailendra Vyakarnam. 2017. *Camels, Tigers and Unicorns: Rethinking Science and Technology Enabled Innovation*. London: World Scientific Press.

Phelan, L., J. McGee, and R. Gordon. 2012. Cooperative Governance: One Pathway to a Stable State Economy. *Environmental Politics* 21 (3): 412–431.

Piggins, Matt P. 2014. *Powering Rural Transformation: Solar PV in the Rural Uganda*. Independent Study Project (ISP) Collection. Paper 1781. http://digitalcollections.sit.edu/isp_collection/1781.

Rogers, Everett. 2003. *Diffusion of Innovations*. 5th ed. Simon and Schuster. isbn:978-0-7432-5823-4.

SACCOL. (2015). Retrieved from http://www.saccol.org.za/what_is_sacco.php.

Twaha, S., M.H. Idris, M. Anwari, and A. Khairuddin. 2012. Applying the Grid-Connected Photovoltaic System as an Alternative Source of Electricity to Supplement Hydropower Instead of Using Diesel in Uganda. *Energy* 37 (1): 185–194.

Wimmer, Nancy. 2012. *Green Energy for a billion poor: How Grameen Shakti Created a Winning Model for Social business*. Vaterstetten: MCRE-Verlag.

6

The Evolving International Gas Market and Energy Security in Nigeria

Tade Oyewunmi

Introduction

This chapter examines the regulatory and institutional dimensions of energy security in Nigeria in the context of an evolving international gas supply market and ongoing domestic petroleum (gas) industry reforms. The global gas market comprises regional markets often grouped based on either the historical patterns of transoceanic shipping, that is, the Atlantic and Pacific Basins, or the primary supra-regions for natural gas trade, that is, North America, Europe, and Asia (Leidos, Inc., for US Energy Information Administration [EIA] 2014; Rogers 2012). The regional and domestic markets have become increasingly international and interconnected mainly due to developments in the liquefied natural gas (LNG) industry and expansion in intraregional gas demand and sup-

T. Oyewunmi (✉)
Centre for Climate, Energy and Environmental Law, UEF Law School, University of Eastern Finland, Joensuu, Finland
e-mail: oyetade.oyewunmi@uef.fi

© The Author(s) 2019
S. Adesola, F. Brennan (eds.), *Energy in Africa*,
https://doi.org/10.1007/978-3-319-91301-8_6

ply mostly within North America, Russia, and Europe, as well as the Asia-Pacific region (BP 2018, pp. 80–83; IEA 2016a, 2017; Rogers 2012). These major markets are served by a capital-intensive network of local and cross-border pipelines, and gas processing and storage facilities. The key stakeholders along the energy value chain for gas supply typically include international and local corporations engaged in production and supply ventures, consumers, and governmental agencies or regulators (Oyewunmi 2018, pp. 14–28; Roggenkamp et al. 2012, pp. 1–10, 413–436; United States Agency for International Development [USAID] 2016; De Vita et al. 2016).

Remarkably, several countries such as Nigeria, Mexico, China, and Egypt are pursuing gas and energy market reforms that are expected to foster private-sector participation and competitiveness, as well as secure more investments across the supply chain. While these market organisation and structural changes occur, recent trends portend more competition for market share and capital investments following developments such as the US shale gas production boom, commercialisation projects in Australia, and significant discoveries in many frontier regions. Given the dictates of behavioural economics and financing in relation to 'risk aversion', in that 'the pain of losing is often considered as greater than the pleasure of winning', especially in a capital-intensive sector like gas supply, countries characterised by regulatory uncertainties and institutional inefficiencies or perceived as high-risk environments are susceptible to 'investment freeze' or avoidance and mothballing by operators, investors, and project developers (Posner 2013; Hu 1990; Bermudez and Pardo 2015). In the evolving contexts of limited capital, dynamic and international market developments, as well as the peculiar attributes of gas and energy which makes it difficult to store and requiring contstant balancing of demand and supply, investors and operators would primarily seek markets with creditworthy buyers and guaranteed reasonable returns on investments. To support firm investment decisions in this regard, the regulatory and institutional aspects of the business environment should foster the liquidity of markets, creditworthiness of buyers as well as reasonable returns on investments. Lack of timely investment decisions in the commercialisation of either associated or non-associated gas, as well as the affiliated supply networks, has security of supply implications, especially in energy markets with an inadequate or unreliable infrastructure (Oyewunmi 2017b, 2018). To be comepetitive, domestic gas markets must develop the necessary organisational institutions—laws, policies, judicial and quasi-judicial decision-makers, independent economic regulators or agencies, and contracts—that can facilitate better adaptation to the demands of the evolving international gas market.

In this chapter, 'energy security' is examined from a 'security of supply' perspective that comprises the legal and institutional condition(s) for safeguarding the reliable access and supply of gas to operating firms, consumers, and stakeholders at reasonable costs, while the risk(s) of significant disruptions are eliminated or adequately mitigated (Oyewunmi 2015a, 2018; Joskow 2007; Von Hirschhausen 2008; Cameron 2007). Given the importance of a reliable and 'secure' gas supply for power generation in the global gas market context, this chapter aims to address the following questions: (i) What are the key institutional features of the typical gas supply to the power industry? (ii) What are the evolving trends and outlook for the international and Nigerian gas supply to the power industry? and (iii) From a law and policy perspective, what are the energy security implications of the trends and outlook for relevant operators, consumers, and stakeholders in Nigeria? Section "The Gas Supply Value Chain" discusses the various elements of the typical gas supply industry, including the nexus between energy security and competitiveness; section "International Gas Markets and Nigeria" examines the evolving international gas markets and the Nigerian scenario; section "Nigerian Gas Supply Industry and Energy Security" focuses on the institutional issues in Nigeria and how they relate to addressing the security of energy supply issues; and section "Conclusion" concludes the chapter.

The Gas Supply Value Chain

The drive towards global gas commercialisation and supply can be attributed to (i) the low-carbon nature of gas compared with other hydrocarbons, such as coal and diesel; (ii) its cost-efficiency and the development of gas-fired power generation technologies, such as combined cycle power plants (CCPPs); (iii) advancements in shipping and long-distance cross-border pipelines enabling access to significant demand centres and remote areas; and (iv) the emergence of the more flexible and competitive LNG supply industry (Sakmar 2015; Stern and Rogers 2011). Another relevant factor has been the liberalisation of the US gas market during the 1980s–1990s, followed by the application of similar paradigms of competition-based gas supply market governance and organisation in the UK and Western Europe, which has led to the increasing commoditisation of gas and more competitive markets (Oyewunmi 2017a). Gas production, processing, storage, and supply is capital intensive and requires a considerable level of technical and operational expertise and system balancing when considered in the context of gas-to-power networks. The physical features of gas, which make it challenging to store, presupposes that investment decisions on production, processing, and supply should take due cognisance of viable demand centres and creditworthy buyers or markets *ex ante*. Thus, adequate and timely investments in production and supply, as well as efficient project planning and management, are critical to energy security, especially in countries such as Nigeria where about 80% of installed electricity generation capacity relies on the gas supply (IEA 2014; World Bank 2004; Santley et al. 2014).

To understand the institutional dynamics of the gas supply industry, it is essential to highlight the legal and contractual nature of property rights and licences, which empowers relevant operators to find, produce, take away, process, and sell gas resources. Historically, upstream petroleum licensing and contracts were tailored towards crude oil exploration and production (E&P), while natural gas discovered in the process (especially associated gas) is often flared, vented, or used in enhanced oil recovery operations. Countries with commercial quantities of natural gas gradually began awarding upstream petroleum rights, licences, or contracts bordering on gas utilisation due to the emergence of significant demand centres and the growth in gas-fired power generation globally (Smith et al. 2010, pp. 1028–1030). The gas supply chain comprises (i) the upstream E&P; (ii) the midstream gas (processing, storage, and transmission); and (iii) the downstream (sales and distribution) segments. The upstream producers hold a licence to explore and produce gas, which is then gathered through small diameter pipelines (gathering lines) from oil and/or gas fields; the gas molecules thereafter go through the processing facilities to remove water and impurities. The gas is compressed to boost its pressure and enable it to flow into large transmission pipelines midstream, which are owned and operated by gas pipeline firms, and then transported to storage, distribution, or marketing centres in the downstream segment (Oyewunmi 2018, 2017b; USAID 2016; Eisen et al. 2015, pp. 539–544).

Upstream Licensing and Contracts

In the US, there is an established framework of absolute or qualified private ownership of oil and gas resources, while in most other countries, the ownership and property rights in oil and gas are vested in the State and managed based on the relevant institutional, legal, and regulatory framework (Aladeitan 2012; Kramer and Anderson 2005). The highly technical and capital-intensive nature of E&P operations connote that governments vested with the property rights would typically engage international oil companies (IOCs) and other private independents as partners, co-venturers, or contractors to develop these resources. In addition, resource-rich countries typically establish national oil companies (NOCs) as the State's commercial participation vehicle in E&P operations (Ledesma 2009). The main forms of international upstream licensing and contractual arrangements include licences and concessions; joint ventures/joint operating agreement (JV/JOA); production sharing contracts (PSCs); and service contracts (i.e., risk service contract [RSC] and pure service contracts) (Naseem and Naseem 2014; Oyewunmi 2018, 2015b; Omorogbe 2003).

(a) **Licences and Concessions**

Old E&P concessions (pre-1960) involved the host government transferring absolute control and ownership of vast areas of land and hydrocarbons within its territorial jurisdiction to IOCs for very lengthy durations, such as 70–90 years. The State was then compensated with payments of royalty and rents, while the IOCs bore all attendant risks and rewards (Smith et al. 2010). Following the formation of the Organization of the

Petroleum Exporting Countries (OPEC) in 1960, and events such as the UN Declaration of Permanent Sovereignty over Natural Resources, 1962 (resolution 1803 (XVII)), OPEC resolution XVI 90 1968, and the New International Economic Order, UN Resolution 3201 of 1974, the major petroleum-producing countries began to participate directly in E&P operations, among other things, with the aim of maximising economic benefits in petroleum resources (Cuervo 2008). Consequently, newer forms of licensing and concessions were introduced, such as Nigeria's oil prospecting licence (OPL) and oil mining lease (OML).[1] Several host governments established NOCs, such as the Nigerian National Petroleum Corporation (NNPC),[2] with the aim of coordinating the State's participation in the industry. Thus, the NOCs could execute JV/JOAs with the IOCs, that is, the former concession holders.

The JV comprises the participation agreement, which defines the relationship and participating interests of the parties, while the JOA defines the legal and operational relationship of the joint venturers by providing for issues such as the operator of the concession, the operating committee, work programme and budget, development or disposition of discovered gas, transfer of participating interests, and so on (Smith et al. 2010). Under the modern concessions and licensing, the State maintains sovereignty over its territory, while incurring capital, commercial, and operational E&P obligations corresponding to the portion of participating interests held by the NOC and as defined in the relevant JV/JOA framework. The IOCs and private companies, on the other hand, are co-venturers with the right to find, produce, and take away oil and gas based on the JV/JOA framework and subject to payment of required royalties and taxes, as provided by the relevant petroleum law(s) of the host country. Due to the considerable capital and technical risk(s) exposure under the licensing or JV/JOA framework, most developing countries now prefer alternative contractual arrangements, such as the PSCs and service contracts.

(b) PSCs and Service Contracts

A PSC is essentially an agreement in which the State holds the licence or lease and appoints the IOC or private E&P company as a contractor to carry out upstream operations. Under the PSC arrangement, the parties agree to share produced oil and gas from the defined contract area in predetermined percentages, following the allocation and payment of relevant tax, royalties, and fees, usually in kind (Smith et al. 2010). The contractor bears all the E&P risks and is generally in charge of operations and the management of the contract area, unless the State party agrees to participate in the venture directly. If no petroleum is found, the contractor typically receives no compensation. By and large, the duration of the E&P period, the evaluation and announcement of a commercial discovery, developing a feasible gas utilisation project, and deciding which party will be primarily responsible for marketing are vital elements relating to gas supply arrangements (Oyewunmi 2015b, 2018).

For instance, in Nigeria, the Petroleum (Drilling and Production) Regulations 1969 ('D&P Regulations') provides that an OPL holder may submit a feasibility study programme or proposal for gas utilisation within five years of the commencement of crude oil production. The Federal Government is empowered under the Petroleum Act (PA) to take produced gas free of charge or at a price without payment of royalty, as well as to approve the price for domestic gas sales. The provisions of the Nigerian Model PSC used in 2005 suggest that more attention is being accorded to gas commercialisation, seemingly in line with the global trend of the 2000s. The 2005 Nigerian Model PSC provides inter alia that when the contractor discovers sufficient gas quantities that could justify commercial development, it shall report the same to the NNPC. The contractor shall investigate and submit proposals for the commercial development while considering local strategic needs to be identified by the NNPC. Both the contractor and NNPC would also execute further gas development agreement(s) that shall recognise the former's right to participate in development projects, the right to recover costs and share in profits. The contractor is also obliged to submit a field development programme to the NNPC.[3]

Likewise, in Tanzania, Article 15 of the 2013 Model Production Sharing Agreement adopted by the Tanzania Petroleum Development Corporation (TPDC) enjoins a contractor who has informed the TPDC of potential commercial interest in discovered natural gas to submit proposals for an appraisal programme within 30 days.[4] Following an approved appraisal programme, the TPDC and the contractor shall execute other agreements on the development, production, processing, and sale of such gas. Such further agreements shall be negotiated in good faith and ensure recovery of all expenses and costs incurred as well as a reasonable return on investments.

Under an RSC, the contractor bears the entire E&P capital and investment risk, while the State retains title to and ownership of the acreage and hydrocarbon in situ. Where the contractor fails to make a commercial discovery, the contract is terminated, with no obligation on either side, but where a commercial discovery is made, the contractor is paid in cash or kind (Smith et al. 2010; Naseem and Naseem 2014). Under a pure service contract, the capital and investment risks are borne primarily by the State, while the contractor is paid a flat fee for its E&P technical services and work carried out. Thus, the contractor is simply a technical service provider working under the State's supervision and has no legal or beneficial interest in the oil and gas resources. Under a technical assistance agreement framework, the IOC or private company is engaged by the host government to provide technical services and technology transfer (Smith et al. 2010). Notably, the RSC and the pure service contracts are common in Latin America and the Middle Eastern countries. Likewise, the PSC, hybrids, and model host government agreements are now more common amongst developing countries, due to the benefits of helping to facilitate the realisation of the underlying objectives of both State and private parties in the most standardised and efficient manner. The Association of International Petroleum Negotiator's (AIPN) suite of model contracts is also commonly adopted in international oil and gas transactions.

In Nigeria, the ownership, property in, and control of all oil and gas resources are vested in the Federal Government, headed by the President. The property and ownership rights are to be administered based on the Laws of the Federation.[5] The PA and D&P regulations constitute the primary legal and regulatory framework for the oil and gas industry (Omorogbe 2003). Through the NNPC, the Federal Government participates and holds majority interests in upstream oil and gas commercial arrangements. The Minister of Petroleum ('Minister') carries out his or her statutory governance and supervisory role as head of the Ministry of Petroleum Resources ('Ministry'), which also includes the Department of Petroleum Resources (DPR) as the industry's primary regulator. The Minister, whose office is part and parcel of the Presidency, and in some cases is actually the President himself where no Minister is appointed, also chairs the board of NNPC in accordance with the NNPC Act. The licensing and regulation of gas supply and pipeline networks is under the purview of the Minister, as stipulated under the Oil Pipelines Act 1956 and the Oil Pipelines Regulations 1995 (the 'Pipelines Regulation') (Oyewunmi 2014; Omorogbe 2003).[6] Until 1992–1993, almost all of Nigeria's upstream operations were carried out under the JV/JOA arrangements between the NNPC and the IOCs or other Nigerian-owned or foreign independents. Since the 1990s, there have been more operations performed through PSCs, while several marginal field licences and sole risk concessions have been issued to indigenous operators (Oyewunmi 2018; Omorogbe 2003).

Centralised and Decentralised Gas Supply Chains

The supply of gas for energy purposes, whether in the form of LNG imported from a gas-producing country or as associated or non-associated gas produced, processed, and transported within a national domestic market, can be carried out in the context of a centralised State-controlled value chain or via a largely decentralised network in which a liberalised and competition-based market structure exists (Peng and Poudineh 2015). From the E&P wellhead to final consumers or large-scale buyers (such as gas-fired electricity generators), the network-bound and natural monopoly nature of the gas supply industry often leads to the development of vertically integrated monopolies or oligopolies, which in some cases have monopsony attributes (Oyewunmi 2018). Such corporations have property and/or commercial interests in gas resources upstream as well as a transmission subsidiary to manage and operate their supply pipelines and ancillary infrastructure. While the actual market structures that exist in the respective countries are mostly hybrids, in a centralised value chain, there is a State-owned or controlled, vertically integrated utility. Such a State-owned or controlled utility is often a subsidiary of the NOC. The utility or gas transportation subsidiary of the NOC owns and operates the entire or most of the domestic gas supply infrastructure within a vertically integrated corporate structure. For instance, the Nigerian Gas Company Limited (NGC) (recently renamed the Nigerian Gas Processing and Transportation Company Limited [NGPTC]) is the NNPC's subsidiary that currently owns and operates the bulk of the domestic transmission and marketing pipeline network in Nigeria.[7] Such centralised gas supply and public utility-styled corporations also existed in countries such as the UK and other EU Member States before the implementation of the US model of liberalisation and economic regulation initiatives which began in the 1990s (Haase and Bressers 2010; Stern and Rogers 2014; Talus 2016).

The decentralised gas supply market archetype is characterised by pro-liberalisation policies such as (i) mandatory or negotiated third-party access (TPA) to the essential supply network of pipelines and supply facilities once controlled or owned by the State-owned utility or private corporations operating as vertically integrated monopolies; (ii) unbundling of network ownership and operation from gas production and sales; (iii) the establishment of an independent economic regulator to efficiently regulate pricing and market access, where natural monopolies exist; and (iv) the emergence of hub markets such as the Henry Hub in the US and the National Balancing Point (NBP) in the UK (Oyewunmi 2017a; Peng and Poudineh 2015). In this regard, a gas producer can execute a purchase and supply agreement with an end-user, such as the operator of a CCPP, and agree on transportation terms with the 'unbundled' pipeline owner or network operating firm, subject to the relevant open access or TPA framework.

Thus, the decentralised paradigms involve multiple private interest holders and corporate participants engaged in non-network segments, such as gas production, import and export, gas storage, sales, and marketing. In such contexts, the transmission and distribution networks could be owned or operated by regulated monopolies, or independent transmission or system operators (Roggenkamp 1997). The primary economic rationale for the competition-based paradigm of decentralised markets is to curtail the propensity of a vertically integrated utility, which may have supply monopoly and upstream monopsony powers to discriminate against customers and third parties. Due to accumulated market power and the absence of competition in the typical centralised contexts, there is also the risk of inadequate commercial motivation to invest efficiently in existing or additional supply infrastructures (Von Hirschhausen 2008; Joskow 1996). Additionally, investment or resource allocation decisions under a centralised or state-controlled framework fraught with regulatory uncertainties and inefficiencies is arguably more susceptible to noncommercial factors such as socio-political considerations, corruption, and bureaucratic bottlenecks, especially in the absence of reasonable and fair competition.

By applying economic regulation and/or antitrust principles, liberalisation should allows multiple producers to gain access to pipelines and supply networks on reasonable, cost-reflective, and non-discriminatory terms, in other words, to sell their volumes on market-led terms, while transmission and system operators are created out of the erstwhile vertically integrated utilities (Baldwin et al. 2012; Spence 2007–2008). Consequently, entry, pricing, and resource allocation in the competitive segments are progressively deregulated, while transmission or distribution network owners are mandated to make their assets available to third parties on non-discriminatory, just, and reasonable terms. An independent economic regulator is also created to ensure accountability, just and efficient, competition-based market interaction. In both the centralised and decentralised archetypes, the role of regulation via formal and organisational institutions is pivotal and can be considered as providing the facilitative means towards realising the policy objectives of competitiveness, security of supply, or sustainability (Oyewunmi 2017a).

Supply Contracts and Organisation

Gas supply arrangements are consolidated following the execution of agreements such as a Gas Supply and Purchase Agreement (GSPA) and Gas Transportation Agreement (GTA). A GSPA or a gas sales agreement between the upstream producer and the supply utility or the pipeline network company aims at securing the former's commitment to sell and the latter's commitment to buy designated quantities of gas to be produced, subject to a predetermined pricing and rate-of-return framework. In some cases, the purchaser, that is, the pipeline network company could be the end-user of the gas in cases where such company also owns or operates a CCPP for gas-fired power generation. Otherwise, the gas purchased is meant for another end-user, such as independent power producers (IPPs), industrial users, and local distributors, or export via LNG facilities and cross-border pipelines. The GTA covers the relevant terms such as transportation tariffs and ancillary service obligations by a pipeline owner and operator for transporting the gas (Oyewunmi 2017b, 2018; Roberts and Maalouf 2014). Traditionally, the agreements have a long-term duration, for example, 20–30 years, with 'creditworthy' buyers and transmission service users; although trends in more competitive markets with the liquidity and financing risks mitigation tools such as in the US and UK trading hubs potray the development of supply contracts with shorter terms. Other significant provisions of these arrangements include a take-or-pay (ToP) clause, a deliver-or-pay (DoP) clause, pricing and price reviews, and destination clauses (Smith et al. 2010).

Concluding relevant terms and the viability of the demand market or buyer's ability to pay is essential to financing and making a final investment decision (FID) on commercialisation projects. Financiers and contracting parties are unlikely to shoulder the significant capital and investments required for these projects, which have long payback periods and involve highly technical and operational requirements, without reasonably firm long-term commitments and efficient risk-allocation mechanisms. In an energy supply context, a long-term contractual framework is often adopted as a tool for ensuring security of demand and security of supply. Generally, it is believed that investments in large-scale, capital-intensive gas processing and transportation facilities or infrastructure will be unfeasible for upstream producers and suppliers without such 'long-term' arrangements in which the terms or production, supply, and purchase are explicitly agreed and supported by a coherent legal and regulatory framework. Nevertheless, the adoption of modern risk-sharing and mitigation contracting tools and market formations that have emerged with liberalisation and short-term hub-based market arrangements in countries such as the US or Western Europe have shown that such risks can be mitigated, so long as functional and efficient institutions are established to provide the required commercial safeguards.

It is worth noting that in the US, the development of competitive gas supply hub trading and spot markets went hand in hand with the creation of a viable natural gas futures market on the New York Mercantile Exchange (NYMEX), which enabled gas buyers and sellers to hedge their price risks and reduce exposure to price volatility (Von Hirschhausen 2008; Eisen et al. 2015). The availability of such financial instruments helped large gas users insure their operations against losses from price volatility attributable to spot-market and short-term arrangements. However, the California energy crisis of 2000–2001 and the Enron Collapse of 2001 are pointers to the severe security of supply problems that could arise even in deregulated markets. In the absence of effective independent economic regulation, which enhances accountability and rule of law, such deregulated markets are equally susceptible to manipulation, opportunism, and rent seeking (Weaver 2004). The development of the competitive gas market in the US in the 1980s–1990s was gradual. The process benefited from an existing and vast network of gas supply infrastructure, as well as strong financial, judicial, and quasi-judicial institutions, and a functional independent regulator, that is, the Federal Energy Regulatory Commission (FERC) (Oyewunmi 2017a). Thus, developing countries such as Nigeria seeking to restructure and develop such competitive and secure markets should note the relevance of underlying institutional factors.

Energy Security and Competitive Gas Supply

Long-term energy security requires timely investments to supply energy in line with economic development and sustainable environmental needs (Barton et al. 2004). Short-term energy security focuses on the ability of the energy system to react promptly to sudden changes in the supply-demand balance (Oyewunmi 2015a). Lack of energy security may, therefore, involve the negative economic and social impacts of either physical unavailability of energy, power outages, and supply disruptions, or prices that are not competitive, unaffordable, or overly volatile. Operators and investors in all commercial ventures, including capital-intensive energy supply, often value the 'risk of losses' more highly than the equivalent 'risk to gains'. Environments characterised by uncertainty (regulatory and commercial), costs, and conflict, or where significant risks cannot be adequately estimated, are therefore typically avoided (Spence 2016–2017; Posner 2013). The general disposition to such environments, as currently exists in the Nigerian petroleum industry due to protracted regulatory reforms and uncertainties, is to 'freeze', divest, or preserve the status quo in the hope of acquiring more information to support future commercial investment decisions.

In the US, for example, the adoption of a public utility and cost-of-service regulation model under the Natural Gas Act 1938, as well as the Supreme Court's decision in *Phillips Petroleum Co. v. Wisconsin* 347 US 672 (1954), was widely perceived as inconsistent with the goal of enhancing competition and an efficient gas market (Anonymous 1982–1983; Oyewunmi 2017a, pp. 245–247). The subsequent application of uncompetitive regulated pricing on the sale of gas at the wellhead (i.e., upstream) led to shortages of gas supply in the midstream interstate pipeline market (Pierce 1995). From the 1980s to 2000s, the relevant US institutions and market operators responded to the unfolding, security of supply challenge with a rigorous process of law-making, contractual and organisational restructuring, judicial decisions by Courts-of-law, and quasi-judicial decisions by the FERC (Eisen et al. 2015, pp. 545–564). Thus, it is expounded that some of the vital benchmarks that enhance gas industry competitiveness and security of supply include (i) legitimacy for the framework or regulatory action; (ii) accountability; (iii) procedural equity and transparency; (iv) expertise of the regulator; and (v) efficiency of the regulatory framework (Oyewunmi 2014, 2018; Baldwin et al. 2012). Regulatory effectiveness in this regard pertains to the capacity of the institutional framework to serve as a means towards realising identified economic and policy objectives at the least possible cost to relevant operators and stakeholders.

International Gas Markets and Nigeria

Global gas supply and trading are mainly carried out via (i) networks of cross-border pipelines and ancillary facilities; and (ii) the LNG supply value chain. The emergence of decentralised and more competitive energy markets in several industrialised and emerging economies is one of the primary drivers of the increasing global trade and commoditisation of gas. Other factors include technological advancements and innovation in gas-fired power generation, pricing and contractual trends in the LNG markets, the shale gas production boom in North America, as well as the commissioning of new LNG and commercialisation projects in countries such as Australia and Russia (International Gas Union [IGU] 2017b;

BP Plc 2016). The BP 2017 Energy Outlook to 2035 reports that LNG trade will grow seven times faster than pipeline gas trade, with LNG accounting for about half of all globally traded gas.[8] Unlike pipeline gas, LNG cargoes can be redirected to different parts of the world in response to regional fluctuations in demand and supply. Thus, previously isolated, domestic or regional gas markets are expected to become more integrated globally (BP 2018; IEA 2017).

Growing demand for flexible LNG supplies can now be met by (i) LNG production volumes that are uncontracted; (ii) volumes that are contracted to a particular destination but redirected; or (iii) contracted volumes open to multiple destinations (which enable gas to flow to demand centres) (IEA 2016b). Additionally, volume per contract has become smaller, reflecting more open markets, more buyers and sellers, and the growing participation of smaller LNG importers. The reliance on oil price indexation as the gas supply pricing mechanism is now diminishing, while the pricing of 'gas' as a commodity on its own right, that is, gas-to-gas pricing, is increasing. Also, the share of contracts with flexible destinations has steadily increased (IEA 2016b). The IGU's global price formation and wholesale market survey reveals that within the past decade:

(a) Adoption of gas-on-gas (GOG) competition in pricing constituted the largest share of total world gas supply/consumption at 45%, predominantly in North America, Europe, the former Soviet Union, and Latin America. The percentage of oil price escalation (OPE) pricing was about 20%. The regulated pricing categories—regulated cost of service (RCS), regulated social and political (RSP), and regulated below cost (RBC)—accounted for about 31%.[9]
(b) The fundamental changes have been the continuous move away from OPE to GOG in Europe; from RBC to RCS, RSP, and GOG in Russia; from RSP to RCS and OPE in China; from RBC to RSP in Iran; and from RBC to RCS in Egypt and Nigeria. GOG and OPE have also recently benefitted from pricing reforms in India and China respectively (IGU 2017a).

The highlighted trends signify a general move towards reforms and developing more competitive price regulation and market governance structures globally. Furthermore, there is a race for market share in global markets by the incumbent and upcoming producers and suppliers. This portends keen competition for scarce investment capital by relevant operators and international firms, including NOCs. Thus, countries, such as Nigeria, which depend heavily on export revenues must develop the necessary regulatory and institutional capacities with the required level of efficiency and responsiveness to address necessary trade-offs and risks arising from demand-side shocks in the global markets (Oyewunmi 2018; IEA 2016b, p. 54). A classic example of how global gas markets are changing and its implications for projected gains for stakeholders in exporting countries such as Nigeria is evident in the upswings and downswings of Nigerian LNG exports to the US between 1999 and 2011 (US EIA 2017).[10] The US shale gas production gained traction in the late 2000s and by 2011 there were clear projections pointing to the US becoming the biggest global gas producer and a net exporter of LNG by 2016 as well as a net exporter of natural gas by 2021 (IEA 2016b, p. 47; IEA 2017; BP 2018). US imports from Nigeria, which peaked at about 95,000 metric cubic feet in 2007, were down to zero by 2012, as pointed out in the US EIA data on LNG imports from Nigeria (US EIA 2017).

In an environment where the NOC and its subsidiary (e.g., Nigeria's NNPC and NGC or NGPTC) are expected to be the primary drivers of investments and infrastructural growth in the domestic market, export revenue losses and non-viable or commercially insecure local markets can lead to significant energy security implications. The conclusion of lingering reforms and enhancing the viability and creditworthiness of operators across the gas-to-power value chain will be essential to the nation's energy security going forward (Oyewunmi 2017a, 2018; Tallapragada 2009; Akinpelu and Iwayemi 2010; Iwayemi 2008).

Nigerian Gas Supply Industry and Energy Security

Considering that Nigeria has about 187 trillion cubic feet of proven gas reserves (the largest in Africa), the nation is consuming only a fraction of what it reasonably could to meet its surging energy demand (BP Plc 2017b; World Bank 2004; Santley et al. 2014). While a limited amount of gas is supplied for power and industrial uses, a more significant portion of gas produced is exported as LNG, flared or reinjected as part of enhanced oil recovery processes (Oyewunmi 2014, 2018; NNPC 2014).

Some of the most significant challenges to the emergence of a competitive and secure gas-to-power value chain in Nigeria revolve around (i) the inadequacy of the domestic gas supply infrastructure, as well as perennial disruptions and affordability of gas supply to power. There is also a perceived lack of creditworthiness and liquidity in the electricity market, which was recently privatised and is undergoing a liberalisation process; (ii) a highly politicised institutional framework that essentially consolidates NNPC/NGC/NGPTC monopoly and control of price regulation and access to pipelines and the market; (iii) a lack of convergence between the evolving electricity market and the gas supply industry; and (iv) rent-seeking and opportunistic private and public stakeholders 'gaming' the regulatory inefficiencies and opacity pertaining to fiscal incentives and resource allocation (Oyewunmi 2017b, 2018; De Vita et al. 2016).

The government's three-in-one role as policymaker, regulator, and commercial operator, via the Minister's office and NNPC, appears directly or indirectly responsible for the regulatory failures and gas market's under-development over the years (Omorogbe 1996; Peng and Poudineh 2017). The drive to reform the legal, organisational, and institutional framework of the oil and gas industry towards international best practices for efficiency, competitiveness, and regulatory effectiveness began in 2000. The broader economic restructuring agenda launched in the 2000s to enhance private-sector participation and liberalisation, and address administrative inefficiencies involved the approval of a National Electric Power Policy 2001 (Electricity Policy), National Energy Policy 2003, and National Oil and Gas Policy 2004 (NOGP). The Electricity Policy was consolidated following the enactment of the Electric Power Sector Reform Act 2005 (EPSR Act) and the creation of the Nigerian Electricity Regulatory Commission (NERC). Furthermore, there was (i) the corporatisation and unbundling of the National Electric Power Authority (NEPA) into 6 generation companies, 11 distribution companies, and a national transmission company; (ii) the privatisation of the successor generation and distribution companies in 2013; and (iii) the declaration of a transitional electricity market towards full liberalisation

(Oyewunmi 2013; Oke 2013). Unfortunately, the restructuring and reforms of the petroleum (oil and gas) industry have remained stunted (Obaseki-Ogunnaike 2017).

The NOGP inter alia prescribed (a) the separation of the Federal Government's regulatory, policy, and commercial roles in the petroleum industry by ensuring distinct institutions perform the respective functions; (b) the corporatisation, restructuring, and eventual privatisation of the NNPC, as well as the unbundling of the NGC into a privatised transmission company, a national gas transport network company, and/or facility management companies; (c) the establishment of a comprehensive National Gas Master Plan (NGMP); (d) the introduction of liberalisation and TPA to the downstream gas sector; (e) creating appropriate gas pricing primarily to facilitate efficiency in the gas supply to power; (f) maintaining a balance between domestic growth and gas export revenue earnings; and (g) enacting a law to consolidate the objectives.

The NGMP provided for a gas pricing policy, the Domestic Gas Supply Obligation (DGSO),[11] and the Gas Infrastructure Blueprint. The attempt to further the NGMP's objectives by issuing the National Domestic Gas Supply and Pricing Policy 2008 (the 'Supply Policy') and the National Domestic Gas Supply and Pricing Regulations 2008 (the 'Supply Regulations') has been incoherent thus far. The Supply Policy's strategic power sector objective of ensuring the delivery of 'low-cost' gas to the power market does not take due cognisance of relevant questions such as the unfeasibility or unavailability of such low-cost gas due to, for example, international gas market dynamics, escalating domestic transaction, and administrative costs, as well as supply disruption issues. The Supply Policy also sought to specify the application of 15% rate-of-return regulation, and other rates and charges. Such policy-based fiscal fixes by the Ministry seem to pre-empt the expected inputs and role of the proposed Department of Gas (DoG) as an independent regulator as well as the extant electricity market regulator, that is, the NERC. The DoG was seemingly established further to the Supply Regulations to function as part of the DPR, which is, in turn, a department under the Ministry. The creation of the Gas Aggregator Company of Nigeria Ltd. (GACN) was also pursuant to the Supply Regulations. The GACN's designated responsibilities seemed to create unnecessary duplication of roles and lack of proper clarity vis-à-vis the DoG, DPR, and NERC concerning gas-to-power regulation (Oyewunmi 2018; Peng and Poudineh 2017).

The objective of establishing an effective independent regulator for either the gas sector and/or the petroleum industry has been hampered by the inability of the government and relevant stakeholders to ensure the enactment of the required reform law or laws, that is, the Petroleum Industry Bill (PIB) of 2008 and 2012. As a result of the political turmoil that followed the legislative process in 2008 and 2012, the current administration decided to split the PIB into four bills, namely, the Petroleum Industry Governance Bill (PIGB), the Fiscal Regime Bill, the Upstream and Midstream Administration Bill, and the Petroleum Revenue Bill (Oyewunmi 2017c). There are also pointers to a possible Petroleum Host Community Bill and a Petroleum Industry Reform Bill (PIRB). The PIGB is the only bill that is currently in circulation and was recently passed by the Senate and House of Representatives. It requires the assent of the President before it becomes law (Oyewunmi 2017c). An administrative re-arrangement or restructuring of NNPC was recently announced in 2016 involving the creation of new subsidiaries, such as Nigerian Gas Processing and Transportation Company Ltd. (NGPTC), Nigerian Gas Marketing Company (NGMC), as well as a gas and power investment division (Oyewunmi 2018).

The IEA 2017 Gas Market Report Series confirms that structural gas shortages reduced the power generation capacity of Nigeria by about 50% in 2015–2016 (IEA 2017). Gas supply shortages and disruptions have paralysed the electricity sector, hampering any new investments in metering, network expansion, and maintenance. Besides the devastating impact of pipeline vandalism in the electricity sector, the country is faced with energy market failure challenges. The DGSO, which was designed to prohibit independent gas producers from exporting gas until they deliver determined volumes for domestic power producers, has been mostly ineffective. For several years, regulated electricity tariffs were set outside of the gas industry's institutional framework, while the pricing and resource allocation in the gas industry were likewise carried out without adequate cognisance to the peculiarities of the power sector's demand and supply dynamics. Thus, pricing and arrangements for a gas supply to power do not often reflect the actual cost(s) of the gas supply as fuel for power generation; resulting in the formation of several ad-hoc inter-ministerial and cross-sectoral committees to move prices from a regulated below-cost towards a cost-of-service regime (IGU 2017a; Oyewunmi 2018, pp. 151–157). Additionally, gas producers prefer to sell gas via the global LNG market than to domestic power producers, while electricity distribution companies are usually unable to pay a cost-reflective and commercially reasonable price for the electricity they buy from the power generation companies (O'Sullivan 2017). In fairness to the newly priva-tised power distribution companies, they seem to have inherited poorly maintained assets and infrastructure, including the required metering and operational networks to effectively determine what would be a fair and reasonable price to charge consumers who themselves are equally wary of private-sector opportunism. Even when the NERC, as the eco-nomic regulator for the power sector, is carrying out its statutory roles alongside the new market operators, there is a constant need to efficiently coordinate with stakeholders and institutions in the oil and gas industry for information regarding the cost, pricing, and availability of fuel, that is, gas.

Ongoing Reforms

Following a review of the current institutional issues, particularly in view of the recent trends in the international oil and gas sector, the current Federal Government approved two policy instruments in the context of a National Economic Recovery & Growth Plan (ERGP 2017–2020) (MPR 2017a, b). These are (i) National Gas Policy 2017 ('Gas Policy') and (ii) National Petroleum Policy 2017 ('Oil Policy'). There is also an ongoing consideration of the draft National Petroleum Fiscal Policy 2016 ('Fiscal Policy'). The new policy initiatives fundamentally reflect the core reformative ideas of the 2004 NOGP, while going a step further to articulate necessary policy revisions for the oil and gas sectors. The Oil Policy reiterates the need for less dependence on oil export revenues and enhancement of economic value from energy resources, especially, by promoting gas-based industrialisation. It proposes the development of a private-sector-led and market-driven industry. Both the Oil Policy and

Gas Policy outline the necessary guidelines for the separation of governments' supervisory, regulatory, and commercial roles within the industry. They prescribe the creation of a single, industry-wide regulatory agency, that is, the Nigerian Petroleum Regulatory Commission, while the Ministry remains responsible for policy directives and supervision. The NNPC is earmarked for restructuring and privatisation once the required laws are enacted, while the administration of the sector is expected to become more transparent and efficient.

The Gas Policy is arguably the first, fully publicised policy framework for promoting gas-based industrialisation and domestic market requirements in Nigeria. It also emphasises the need for maintaining a significant presence in international markets. On the organisational structure for the industry, the Gas Policy recommends (i) a mix of public-private participation; (ii) restructuring of NGC into separate transport and gas marketing companies; (iii) strategic partnerships to support the operations of the NGPTC; (iv) developing wholesale market competition; and (v) implementation of the DGSO and reviewing the future role of the GACN. With regard to pricing reforms, the Gas Policy stipulates that the upstream gas price for domestic sales will be set by netback from export-parity prices. It proposes a transitional period after which a market-led wholesale gas pricing will be the norm. In consonance with the Fiscal Policy, the gas industry operators should also expect a fiscal framework which recognises gas as a standalone commodity and industry, separate from oil.

While the Gas Policy's statement that private operators must view the DGSO framework as their own contribution to national development and doing business in Nigeria seems understandable from a government perspective, it should be highlighted that the idea of grounding the issuance and renewal of upstream licences on compliance with the DGSO may be unrealistic and counter-intuitive. Note that NNPC and IOCs hold the most resourceful E&P acreages under various arrangements, such as JV/JOA or PSCs. Under PSCs, NNPC is the licence holder or leaseholder, while the corporation is also a leaseholder to the extent of participating interests held in JV/JOAs. Thus, it is unrealistic to suggest that NNPC will be denied the renewal or issuance of a licence or OML. Arguably, the failure to treat IOCs and NNPC equally without discrimination in this regard may have international law and investment protection implications (Hirsch 2011; Cameron 2014).

Another possible implementation challenge with the Gas Policy relates to its prescription of a National Gas Focal Point and dedicated project desks within the Ministry. The Gas Focal Point is expected to carry out oversight and functional implementation roles for overcoming any obstacles and ensuring consensus and a coordinated development among all industry participants. The 'Project Desks' will serve as the interface between project developers and government agencies. It should be noted that if law or regulation does not clearly define the role and scope of authority of these offices within the Ministry, this may lead to overlaps and administrative bottlenecks vis-à-vis the expected role of the proposed regulator. Such a regulator, with potential economic and quasi-judicial functions, should be allowed and equipped to act independently and competently. Overall, it is also noted that national plans, policies, and guidelines issued by one administration can be replaced and changed by the next or the same administration. Therefore, unless the government takes more law-based and firm implementation steps, the atmosphere of regulatory uncertainties and inefficiency perceptions by current and potential investors in the gas supply for domestic energy uses could continue to undermine energy security in Nigeria.

Conclusion

The capital-intensive, commercial, and operational elements of gas supply projects mean that they may remain unfeasible if only dedicated to an under-developed or non-viable domestic energy market. Thus, developing a competitive, secure, and reliable local supply value chain, while maintaining global export-related capacities, is essential to energy security in countries such as Nigeria.

In promoting the security of supply dimensions of energy security, the instrumental role of regulation through formal institutions such as laws, judicial and quasi-judicial decisions, as well as public and private organisational institutions such as contracts and independent regulators, cannot be over-emphasised. Regardless of the approach, the objectives involve creating a stable investment climate to underpin significant investments, usually spanning decades, in the country's petroleum resources. An effective gas sector policy and institutional framework would comprise the promotion of gas deliverability; affordability of gas; commercialisation of supply to enable secure willing buyer/willing seller arrangements; availability of gas to meet energy demand and supply requirements; competitive and non-discriminatory market access, and clearly defined regulations, which promote transparency and role clarity amongst stakeholders; and cover issues such as third-party access, pipeline ownership, and tariff structures.

Acknowledgement The author is grateful to the Fortum Foundation of Finland, the Finnish Cultural Foundation, and the UEF Law School's Academy of Finland Project 276974 on "the impact of shale gas in EU energy law and policy" for grants and/or contributions which supported the research work carried out while writing this chapter.

Notes

1. The Petroleum Act 1969 CAP P10 Laws of the Federation of Nigeria 2004 ("PA").
2. The NNPC was created pursuant to the Nigerian National Petroleum Corporation Act 1977.
3. Author's copy. See also OGEL legal and regulatory database at www. ogel.org/legal-and-regulatory-countries-browse.asp?country=156.
4. See the OGEL Journal's Legal and Regulatory database collection on Tanzania at www.ogel.org/legal-and-regulatory-detail.asp?key=11506. Accessed August 5, 2017.
5. The Constitution of the Federal Republic of Nigeria 1999 as amended.
6. CAP P13, Laws of the Federation of Nigeria, 2004.
7. The current domestic gas pipeline infrastructure mainly comprises two unintegrated pipeline networks of approximately 1100 kilometres: (i) the Alakiri-Obigbo–Ikot Abasi Pipeline (the Eastern Network), and (ii) the Escravos–Lagos Pipeline System (ELPS) (the Western Network), as well as the dedicated pipeline infrastructure owned by the Nigerian Liquefied Natural Gas Company (NLNG), the NNPC/SPDC/Total JV, and the Chevron/NNPC JV.

8. See BP Plc. 2017. BP Statistical Review of World Energy 2017, p. 35 available at www.bp.com/en/global/corporate/energy-economics/statis-tical-review-of-world-energy.html. Accessed October 12, 2017; BP Plc. 2017. BP Energy Outlook to 2035 (2017 edition) p. 56, available at www.bp.com/content/dam/bp/pdf/energy-economics/energyout-look-2017/bp-energy-outlook-2017.pdf or www.bp.com/en/global/cor-porate/energy-economics/energy-outlook/energy-outlook-downloads.html. Accessed October 12, 2017.

9. The categories of OPE, GOG, Bilateral Monopoly, and Netback from Final Product can be broadly described as "market-based" pricing, while the categories of RCS, RSP, and RBC can be classified as "regulated" pricing.

10. See data from US Energy Information Administration on the U.S. Liquefied Natural Gas Imports from Nigeria (Million Cubic Feet) (1997–2016) at https://www.eia.gov/dnav/ng/hist/n9103ng2a.htm. Accessed October 12, 2017.

11. The DPR's 2014 National Oil and Gas Report, reveals a dismal compli-ance level, that is, an annual 20%–35% average compliance with DGSO requirements between 2008 and 2014. The major reason for poor indus-try compliance is the preference of producers for the more competitively priced export market; inadequate domestic pipeline infrastructure; slip-pages in project execution and budget constraints; failure of swap deals; and non-readiness of offtake power plants.

References

Akinpelu, L.O, and Akin Iwayemi. 2010. *Appropriate Gas Price Determination in the Emerging Nigerian Gas Market*. Tinapa – Calabar, Nigeria, Society of Petroleum Engineers, 31 July–7 August. www.onepetro.org/conference-paper/SPE-136959-MS. Accessed 15 June 2016.

Aladeitan, Lanre. 2012. Ownership and Control of Oil, Gas, and Mineral Resources in Nigeria: Between Legality and Legitimacy. *Thurgood Marshall Law Review* 38: 159–198.

Anonymous. 1982–1983. Natural Gas Regulation and Market Disorder. *Tulsa Law Journal* 18: 619–648.

Baldwin, Robert, Martin Cave, and Martin Lodge. 2012. *Understanding Regulation: Theory, Strategy, and Practice*. 2nd ed. Oxford/New York: Oxford University Press.

Barton, Barry, Catherine Redgwell, Anita Rønne, and Donald N. Zillman. 2004. Energy Security in the Twenty-First Century. In *Energy Security: Managing Risk in a Dynamic Legal and Regulatory Environment*, ed. Barry Barton, Catherine Redgwell, Anita Rønne, and Donald N. Zillman, 457–470. Oxford: Oxford University Press.

Bermudez, Jose Luis, and Michael S. Pardo. 2015. Risk, Uncertainty, and Super-Risk. *Notre Dame Journal of Law, Ethics & Public Policy* 29: 471–496.

BP Plc. 2016. *Energy Outlook To 2035 (2016 Edition)*. London: BP Plc.

———. 2017a. *BP Statistical Review of World Energy, June 2017*. London: BP Plc.

———. 2017b. *Energy Outlook to 2035 (2017 Edition)*. London: BP Plc.

———. 2018. *Energy Outlook to 2040 (2018 Edition)*. London: BP Plc.

Cameron, Peter. 2007. *Competition in Energy Markets Law and Regulation in the European Union*. 2nd ed. Oxford/New York: Oxford University Press.

———. 2014. In Search of Investment Stability. In *Research Handbook on International Energy Law*, ed. Kim Talus, 124–147. Cheltenham: Edward Elgar.

Cuervo, Luis E. 2008. OPEC: From Myth to Reality. *Houston Journal of International Law* 30 (2): 433–615.

De Vita, Glauco, Oluwatosin Lagoke, and Sola Adesola. 2016. Nigerian Oil and Gas Industry Local Content Development: A Stakeholder Analysis. *Public Policy and Administration* 31 (1): 51–79.

Eisen, Joel, Emily Hammond, Jim Rossi, David Spence, Jacqueline Weaver, and Hannah Wiseman. 2015. *Energy, Economics and the Environment, Cases and Materials*, University Casebook Series. 4th ed. St. Paul: Foundation Press.

Haase, Nadine, and Hans Bressers. 2010. New Market Designs and their Effect on Economic Performance in European Union's Natural Gas Markets. *Competition and Regulation in Network Industries* 11 (2): 176–206.

Hirsch, Moshe. 2011. Between Fair and Equitable Treatment and Stabilization Clause: Stable Legal Environment and Regulatory Change in International Investment Law. *Journal of World Investment & Trade* 12: 783–806.

Hu, Henry. 1990. Risk, Time, and Fiduciary Principles in Corporate Investment. *UCLA Law Review* 38 (2): 277–390.

International Energy Agency (IEA). 2014. *African Energy Outlook: A Focus on Prospects in Sub-Saharan Africa*. Paris: IEA Publications.

———. 2016a. Key Natural Gas Trends. In *Natural Gas Information (2016 edition)*, vii–xi. Paris: IEA Publications.

———. 2016b. *Global Gas Security Review 2016*. Paris: IEA Publications.

———. 2017. *Market Report Series: Gas 2017: Analysis and forecasts to 2022*. Paris: IEA Publications.

International Gas Union (IGU). 2017a. *2016 IGU Wholesale Gas Price Survey.* IGU Publications.

————. 2017b. *IGU World LNG Report – 2017 Edition.* IGU Publications.

Iwayemi, Akin. 2008. Nigeria's Dual Energy Problems: Policy Issues and Challenges. *International Association for Energy Economics Energy Forum,* Fourth Quarter 53: 17–21.

Joskow, Paul. 1996. Introducing Competition into Regulated Network Industries: From Hierarchies to Markets in Electricity. *Industrial and Corporate Change* 5 (2): 341–382.

————. 2007. Supply Security in Competitive Electricity and Natural Gas Markets. In *Utility Regulation in Competitive Markets Problems and Progress,* ed. Colin Robinson. Cheltenham: Edward Elgar.

Kramer, Bruce, and Owen Anderson. 2005. The Rule of Capture – An Oil and Gas Perspective. *Environmental Law* 35 (4): 899–954.

Ledesma, David. 2009. *The Changing Relationship between NOCs and IOCs in the LNG Chain.* NG 32. Oxford Institute for Energy Studies.

Leidos, Inc., for US Energy Information Administration (EIA). 2014. *An Introduction to Global Natural Gas Markets, Drivers, and Theory.* US EIA. www.eia.gov/workingpapers/pdf/global_gas.pdf. Accessed 14 Feb 2016.

Ministry of Petroleum Resources (MPR), Nigeria. 2017a. *National Gas Policy.* MPR.

————. 2017b. *National Petroleum Policy.* MPR.

Naseem, Mohd, and Saman Naseem. 2014. World Petroleum Regimes. In *Research Handbook on International Energy Law,* ed. Kim Talus, 149–180. Cheltenham: Edward Elgar.

Nigerian National Petroleum Corporation (NNPC). 2014. *Annual Statistical Bulletin (second edition),* last modified on January 1, 2016: 1–47 at p. 34. www.nnpcgroup.com/Portals/0/Monthly%20Performance/2014%20 ASB%202nd%20Edition.pdf. Accessed 12 Feb 2016.

O'Sullivan, Kyran. 2017. *Nigeria – Nigeria Electricity and Gas Improvement Project (NEGIP) P106172 – Implementation Status Results Report.* Washington, DC: World Bank Group.

Obaseki-Ogunnaike, Donna. 2017. Understanding the Nigerian Petroleum Industry Governance Bill 2016. *OGEL Journal 1 (2017)* Special Issue on Oil and Gas Law and Policy in West Africa.

Oke, Yemi. 2013. *Nigerian Electricity Law and Regulation.* Lagos: The Law Lords Publications.

Omorogbe, Yinka. 1996. Law and Investor Protection in the Nigerian Natural Gas Industry. *Journal of Energy & Natural Resources Law* 14: 179–192.

———. 2003. *Oil and Gas Law in Nigeria*. Lagos: Malthouse Press.

Oyewunmi, Tade. 2013. International Best Practices and Participation in a Private Sector Driven Electricity Industry in Nigeria: Recent Regulatory Developments. *International Energy Law Review* (8): 306–314.

———. 2014. Examining the Legal and Regulatory Framework for Domestic Gas Utilization and Power Generation in Nigeria. *Journal of World Energy Law & Business* 7 (6): 538–557.

———. 2015a. Energy Security and Gas Supply Regulation in the European Union's Internal Market. *European Networks Law & Regulation Quarterly* 3 (3): 187–202.

———. 2015b. Natural Gas Exploration and Production in Nigeria and Mozambique: Legal and Contractual Issues. *OGEL 1 (2015),* Special Issue on Natural Gas Developments: An International and Challenging Legal Framework.

———. 2017a. Examining the Role of Regulation in Restructuring and Development of Gas Supply Markets in the United States and the European Union. *Houston Journal of International Law* 40 (1): 191–296.

———. 2017b. Regulatory and Policy Issues for Natural Gas Supply to Power Markets: Examining the Energy Supply Crisis in Nigeria. *OGEL Journal* 1 (2017). Special Issue on Oil and Gas Law and Policy in West Africa. www.ogel.org/article.asp?key=3677. Accessed 21 May 2017.

———. 2017c. Nigeria – Energy Policy – (Sub-Saharan Africa – Energy Policy Section). In *Encyclopaedia of Mineral and Energy Policy*, ed. Günter Tiess, Tapan Majumder and Peter Cameron. Springer. https://doi.org/10.1007/978-3-642-40871-7_158-1. Accessed 12 Nov 2017.

———. 2018. *Regulating Gas Supply to Power Markets: Transnational Approaches to Competitiveness and Security of Supply*. Energy and Environmental Law and Policy Series. Kluwer Law International.

Peng, Donna, and Rahmat Poudineh. 2015. *A Holistic Framework for the Study of Interdependence Between Electricity and Gas Sectors*. EL 16. Oxford Institute for Energy Studies.

———. 2017. *Gas-to-Power Supply Chains in Developing Countries: Comparative Case Studies of Nigeria and Bangladesh*. EL 24. Oxford Institute for Energy Studies.

Pierce, Richard J. 1995. The Evolution of Natural Gas Regulatory Policy. *Natural Resources & Environment* 10 (1): 53–85.

Posner, Richard. 2013. Behavioral Finance Before Kahneman. *Loyola University Chicago Law Journal* 44: 1341–1347.

Roberts, Peter, and Ruchdi Maalouf. 2014. Contractual Issues in the International Gas Trade: LNG- the Key to the Golden Age of Gas. In *Research Handbook on International Energy Law*, ed. Kim Talus, 329–357. Cheltenham: Edward Elgar.

Rogers, Howard. 2012. *Impact of a Globalising Market on Future European Gas Supply and Pricing: The Importance of Asian Demand and North American Supply*, NG 59. Oxford: Oxford Institute for Energy Studies.

Roggenkamp, Martha M. 1997. Implications of Privatisation, Liberalisation and Integration of Network bound Energy Systems. *Journal of Energy & Natural Resources Law* 15 (1): 51–61.

Roggenkamp, Martha M., Lila Barrera-Hernández, Donald N. Zillman, and Iñigo del Guayo, eds. 2012. *Energy Networks and the Law: Innovative Solutions in Changing Markets: Innovative Solutions in Changing Markets*. Oxford: Oxford University Press.

Sakmar, Susan. 2015. *Energy for the 21st Century: Opportunities and Challenges for Liquefied Natural Gas (LNG)*. Cheltenham: Edward Elgar.

Santley, David, Robert Schlotterer, and Anton Eberhard. 2014. *Harnessing African Natural Gas: A New Opportunity for Africa's Energy Agenda?* 89, 622. Washington, DC: World Bank Group. http://hdl.handle.net/10986/20685. Accessed 20 June 2015.

Smith, Ernest E., John S. Dzienkowski, Owen L. Anderson, John S. Lowe, Bruce M. Kramer, and Jacqueline L. Weaver. 2010. *International Petroleum Transactions*. 3rd ed. Westminster: Rocky Mountain Mineral Law Foundation.

Spence, David. 2007–2008. Can Law Manage Competitive Energy Markets. *Cornell Law Review* 93: 765–818.

———. 2016–2017. Naive Energy Markets. *Notre Dame Law Review* 92: 973–1030.

Stern, Jonathan, and Howard Rogers. 2011. *The Transition to Hub-Based Gas Pricing in Continental Europe*. NG 49. Oxford: Oxford Institute for Energy Studies.

———. 2014. *Dynamics of a Liberalised European Gas Market – Key Determinants of Hub Prices, and Roles and Risks of Major Players*. NG 94. Oxford: Oxford Institute for Energy Studies.

Tallapragada, Prasad V.S.N. 2009. Nigeria's Electricity Sector- Electricity and Gas Pricing Barriers. *International Association for Energy Economics Energy Forum*, First Quarter: 29–34.

Talus, Kim. 2016. *Introduction to EU Energy Law*. Oxford: Oxford University Press.

United States Agency for International Development (USAID). 2016. *Understanding Natural Gas and LNG Options.* USAID/Us Department of Energy/Us Energy Association.

US Energy Information Administration (US EIA). 2017. U.S. Liquefied Natural Gas Imports from Nigeria (Million Cubic Feet) (1997 to 2016). Release Date January 31, 2018. https://www.eia.gov/dnav/ng/hist/n9103ng2a.htm. Accessed 5 Feb 2018.

Von Hirschhausen, Christian. 2008. Infrastructure, Regulation, Investment and Security of Supply: A Case Study of the Restructured US Natural Gas Market. *Utilities Policy* 16 (1): 1–10.

Weaver, Jacqueline L. 2004. Can Energy Markets be Trusted? The Effect of the Rise and Fall of Enron on Energy Markets. *Houston Business and Tax Law Journal* 4: 1–151.

World Bank. 2004. *Nigeria Strategic Gas Plan,* ESM279. Washington DC: IBRD/World Bank. https://openknowledge.worldbank.org/handle/10986/19892. Accessed 15 July 2015.

Part III

Energy Transition: Clean Technology, Sustainability, and Affordability

7

Clean Technologies and Innovation in Energy

Tosin Somorin, Ayodeji Sowale,
Adefolakemi Serifat Ayodele, Mobolaji Shemfe,
and Athanasios Kolios

Introduction

African economies are faced with diverse challenges—the dilemma of increasing food, water, energy, and sanitation services to millions of people who currently live without access to basic services and infrastructures, and the need to operate low-carbon systems that meets environmental standards. The problems of poor access to modern energy services are

T. Somorin (✉) • A. Sowale • A. Kolios
Cranfield University, Cranfield, UK
e-mail: t.o.onabanjo@cranfield.ac.uk; A.O.Sowale@cranfield.ac.uk;
a.kolios@cranfield.ac.uk

M. Shemfe
University of Surrey, Guildford, UK
e-mail: m.shemfe@surrey.ac.uk

A. S. Ayodele
University of Ibadan, Ibadan, Nigeria

© The Author(s) 2019
S. Adesola, F. Brennan (eds.), *Energy in Africa*,
https://doi.org/10.1007/978-3-319-91301-8_7

more prevalent in the least developed countries (LDCs) and Sub-Saharan Africa (SSA) (Legros et al. 2009). Currently, the demand for energy in Africa surpasses supply; almost 30 out of 54 countries suffer from erratic electric power supply (IEA 2014). In concrete terms, about 621 million people (about two in three persons of the total population) are without access to electricity (Africa Progress Panel 2015). It is also estimated that 730 million Africans still bank on the traditional use of biomass as their primary source of power supply (IEA 2014). Overall, the continent's average electric power consumption was approximated at 800 kWh per capita in 2014 and about 200 kWh in SSA (IEA 2017). This value is significantly low compared to those of North America, Europe, and Central Asia that are in excess of 5000 kWh per capita for the same period (Masud et al. 2007). With a population expected to grow to be above 2.54 billion by 2050 in Africa (PRB 2017)—more than half of the projected global population increase—the continent must harness the ample number of renewable energy sources, with which it is endowed, to meet its energy demands. This requires substantial advances in energy innovations and technologies.

Clean energy technologies can stimulate economic growth and development, and provide several environmental advantages, such as reduced waste quantities, fossil fuel consumption, and climate benefits (EPA 2011). These technologies can vary from small, medium to large scales and are applied in various settings. At household levels, they can be used for meeting energy needs for cooking, lighting, heating, cooling, and so on. In a commercial or industrial environment, they can include energy solutions for production, transportation, storage, and so on. Essentially, the technological solution should have negligible environmental impact and provide some form of socio-economic benefits, such as employment, revenue creation, and cost-competitiveness (Shen and Power 2017), although its primary function is to meet direct or indirect energy needs. This chapter therefore provides an overview of the clean energy sources and their potential in Africa. The progress in clean energy technologies is discussed with considerations and focus on proven, low-carbon, and environmentally friendly energy systems on small to industrial scales. This includes various applications and the criteria for developing similar solutions. The chapter not only discusses the challenges and barriers to the development, transfer, and diffusion of innovative energy technologies in Africa, but also highlights some mitigation strategies to overcome them.

Clean Energy Sources and Potentials

The energy sources in Africa are numerous and diverse—ranging from fossil fuels to renewable resources, such as solar, hydro, wind, biomass, and geothermal (Iwayemi 1998). The renewable supplies are abundant, owing to Africa's favourable geographical location and weather conditions. There are accounts of large distributions of solar energy in the northern part of the continent, areas with the highest radiation levels in the world (Ramachandran et al. 2009). Wind energy is widely available across the continent, especially in the northern regions (IRENA 2012). Resources, such as biomass and hydro energy, are abundant in the wet and forest-dense southern and central regions of Africa (Fig. 7.1). Despite this abundance, 50% of the continent's primary energy supply is derived from fossil fuels, mainly oil, coal, and natural gas. Energy sources, such as geothermal, solar, and hydro, account for less than 1% and this is an indication of the underutilisation of clean energy sources.

Currently, wind energy is used on small to medium scales for powering pumps and driving electrical generators and other machinery while on a large scale, it is used for power production. According to data by the International Renewable Energy Agency (IRENA), South Africa has the highest installed capacity for wind energy in Africa with 1473 MW in 2016, accounting for about 40% of wind energy capacity in Africa (IRENA 2017). This is followed by Morocco (798 MW) and Egypt (750 MW). Countries such as Ethiopia are rapidly developing the exploitation of wind energy capacity (324 MW as at 2016), while Nigeria is yet to harness its wind energy potential with a meagre capacity of 3 MW (IRENA 2017). Hydro resources are typically employed for irrigation and electricity generation. Extensive hydro reserves, such as the Nile, Zambezi, Niger, and Congo rivers, are harnessed from multiple points and on a large scale, usually with a storage dam for electricity production of up to 500 MW or on a small scale with or without a dam for electricity

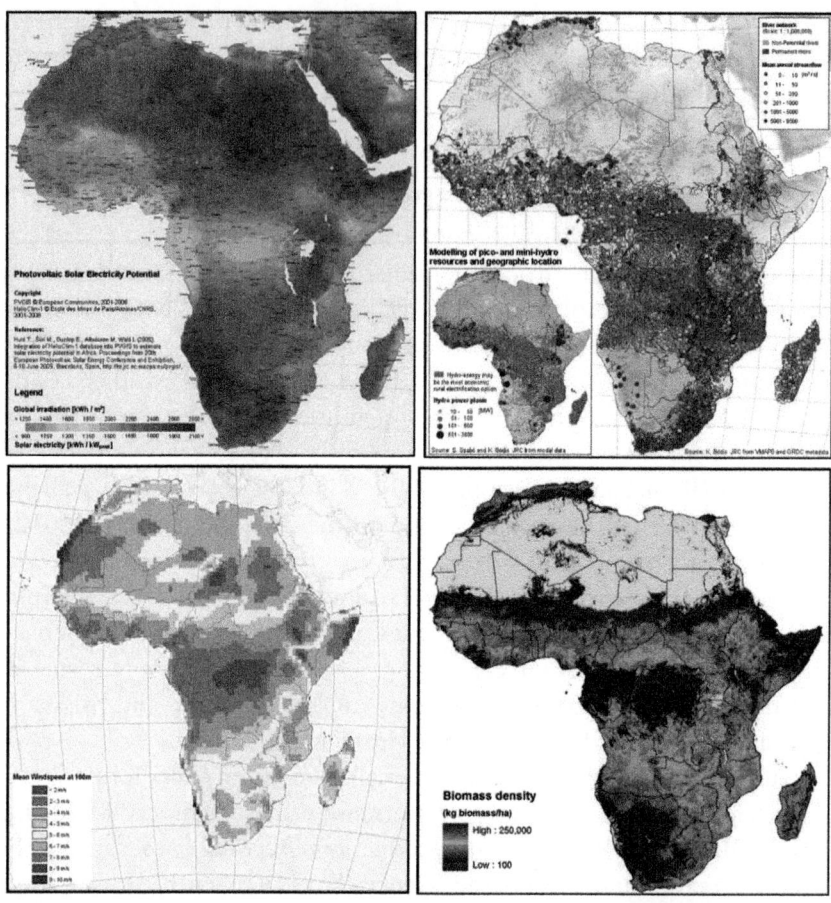

Fig. 7.1 Renewable energy potential in Africa: Solar irradiation (kWh/m²) mapping of Africa—*Top left*, River network in Africa in terms of mean annual streamflow (m³/s). Sub-figure shows areas where mini hydro may be the most economic electrification option—*Top right*, Geographical distribution of average wind speed over Africa at 100 m height—*Bottom left*, Biomass (woody and agricultural) density across Africa—*Bottom right*. (Source: Belward et al. 2011)

production between 5 kW and 10 MW (Eberhard et al. 2016). Geothermal energy is mainly employed directly in industrial processes for low-temperature heat operation, especially in the manufacturing industry, as it is relatively inexpensive and can reduce the use of fossil fuels. Along the Rift valley, an area that covers Mozambique to Djibouti, geothermal energy is reported to produce electricity of about 15 GW (United States Energy Association 2017). Kenya was reported to be the first African country to exploit geothermal energy with a generating capacity of 1091 MW (IRENA 2017).

Biomass is unique among other clean energy sources because of its wide range of use and application; it can be used for the generation of heat, electricity, chemicals, and fuels (Shemfe et al. 2015; Shemfe 2016; Somorin and Kolios 2017). It includes resources such as wood fuel, forest residues, and agricultural wastes that are typically used as fuel for cooking and heating (IEA 2006). Biomass also includes energy crops that are increasingly being used for liquid fuels in small and medium scale industries, and animal wastes from farm animals (HSUS 2009). The agricultural residues include stalks, straw, husks, leaves, branches from plants, and by-products of agricultural processes (UN DESA 2007). Animal wastes, also known as manure, are pure or mixtures of animal excreta and bedding materials. Wood fuels are derived from forest trees and wood processing industries, for example, sawmills. One-fifth of these resources are converted into charcoal for cooking, a process that contributes to deforestation (Atteridge et al. 2013). Although, there are alternatives such as kerosene and liquefied petroleum gas (LPG), wood fuel is the preferred option because of its wide availability and low cost (IEA 2006). Its inefficient use is, however, associated with indoor air pollution, and reported as damaging to health and the environment (Barría 2016; Fullerton et al. 2008). As such, more reliable and productive end use of traditional biomass, particularly wood fuel, is essential towards the transition to sustainable energy supply. In Africa, liquid fuels (e.g. ethanol and biodiesel) are typically derived from food crops such as sugarcane, molasses, sunflower, and soybean, although they can be derived from lignocellulosic biomass (Balan 2014). According to IRENA (2014), ethanol production is promising in the eastern and southern parts of Africa—65% production potential in southern Africa and 20% in East and Central Africa. In West Africa, particularly in Nigeria, Ghana, and Benin, oil palm is widely produced, but not essentially for fuel. However, there has been considerable increase in investments to secure land for biofuel production. This practice is considered to be in competition with food and thus leads to land shortages, particularly when the fuel is derived from a food crop

Table 7.1 Technical energy potential for different energy sources and African regions

Region	CSP (TWh/yr)	PV (TWh/yr)	Wind (TWh/yr)	Hydro (GWh/yr)
Central Africa	29,909	61,643	12,395	59,393
Eastern Africa	175,777	219,481	165,873	101,492
Northern Africa	93,544	109,033	130,316	570,730
Southern Africa	149,610	162,817	108,235	334,600
Western Africa	22,747	103,754	40,846	415,857
Total (Africa)	**471,587**	**656,728**	**457,665**	**1,482,072**

Source: Hermann et al. (2014), IRENA (2014)

(Tenenbaum 2008). The production of oil from lignocellulosic biomass is however yet to be widely commercialised because of its high energy and resource requirements (Balan 2014). In these instances, efficient energy conversion processes and technologies can play a huge role (Table 7.1).

Using a Geographical Information System (GIS) computational approach, the annual theoretical energy potential in Africa is estimated at 470 and 660 Petawatt hours (PWh) for solar energy based on Concentrated Solar Power (CSP) and Photovoltaic (PV) technologies (Hermann et al. 2014). Wind energy is estimated at 460 PWh (Hermann et al. 2014) and 1,584,670 Gigawatt hours (GWh) per year (IRENA 2015a) for hydropower potential. Although these theoretical energy potentials are maximum values, with no consideration of conversion efficiencies and losses, they are considered more than sufficient to meet Africa's future energy demand, if harnessed properly. Based on IRENA (2015a) projections, the use of modern renewable technologies across African countries and their economic sectors can contribute significantly to the total final energy consumption by 2030. There are however technological constraints, such as poor availability, accessibility, reliability, and affordability of energy technologies and other factors, that limit the development, transfer and diffusion of new and innovative energy technologies, and provision of energy services in Africa. The UNEP and EPO Report (2013) suggested the importance of developing appropriate and tailored energy technologies for the different regions of Africa, since clean energy sources are diverse in quantity and quality. Wu et al. (2017) emphasised the importance of integrated designs and energy solutions with an adequate and well-structured strategy for harnessing renewable energy sources in Africa.

For instance, the study proposed the development of wind farms and solar power plants, coupled with the international sharing of power to limit the construction of conventional power plants.

The next section presents the progress in the development and provision of clean energy technologies in Africa and for different applications on small, medium, and large industrial scales.

Clean Energy Technologies in Africa

The plethora of renewable energy resources in Africa can be tapped and converted into useful work, chemicals, and other forms of energy via various conversion technologies. Although financial and technical barriers have been historically and partly responsible for the slow growth of renewable energy technologies in Africa, the main culprit has been limited interest from policymakers to transition to these technologies. Nevertheless, there has been a recent renaissance to power Africa's economic growth with its numerous clean energy resources. In recent times, there has been increasing demand for clean energy supply in critical economic sectors, including agriculture, power, and manufacturing (World Bank 2017a, b). As such, Africa's economy is projected to increase to 4% in the next half decade, in terms of average annual Gross Domestic Product. This resurgence was further strengthened by the 2015 Paris Agreement with the goal to limit the rise in mean global temperature to 2 °C in order to mitigate the effects of global climate change. Projections of the impact of climate change on Africa look dire, as many vulnerable populations are bound to be affected by the consequential droughts and ecosystem degradation, albeit the continent contributes the lowest greenhouse gas (GHG) emissions (about 4%) to the global total (Sy 2015). Thus, African countries would have to adapt to these impending changes in the climate. Some mechanisms are already in place, while others are still in the pipeline to apply ecosystem-based adaptations in the agriculture sector in combination with clean energy technologies towards securing food supply, enhancing economic prosperity, reducing GHG emissions, and conserving the ecosystem. Overall, it is essential for African countries to adopt a near-sizable mix of clean energy technologies to supplement conventional technologies, and by so doing expand access to both centralised and decentralised solutions.

In the following subsections, existing and plausible clean energy technologies for harnessing Africa's clean energy resources are elaborated.

Solar Energy Technologies

Solar energy, in the form of light and/or heat, can be harnessed and converted into electricity, mechanical work, and/or heat energy via the use of CSP, PV, and other hybrid solar technologies. On small (domestic) and large scales, this energy is used for cooking, heating, water purification, lighting and cooling, mini-grid development and rural electrification, as well as commercial and industrial processes (Feron 2016).

In many parts of SSA, small- to mid-scale solar projects are gaining ground. For example, an 8.5 MW solar farm was developed in Rwanda in 2014 with 2800 solar panels. PV systems are rapidly being deployed in countries, such as Ghana, Kenya, Namibia, South Africa, and Zimbabwe, due to the relatively low cost of PV systems (IRENA 2016). Currently, residential rooftop systems account for about 90% of the continent's PV market, but large-scale PV systems are nearly non-existent (Jones 2017; Suberu et al. 2013). The use of CSP is also particularly low, only accounting for less than 5% of solar power generation. Monumental solar projects are however underway in North Africa, such as the Noor solar power complex in Morocco. These projects are driven by the feasibility of transmission of high-voltage direct current to consumption centres in Europe. Thus, there is opportunity for the continent to fully harness its solar energy resources via the proliferation of solar energy technologies. Table 7.2 shows the distribution of solar PV and the use of solar-powered water heaters in selected SSA countries.

Hydropower Technologies

Hydropower is generated from the turning of turbines by the kinetic energy of fast moving and falling water bodies; the resulting electricity is clean and emission free. The report on the Southern African Power Pool

Table 7.2 Distribution of PV and domestic solar water heater installed capacity in selected Sub-Saharan African countries

Country	
	Estimated number of systems (kW$_p$)
Uganda	538 (152)
Botswana	5724 (286)
Zambia	5000 (400)
Zimbabwe	84,468 (1689)
Kenya	150,000 (3600)
South Africa	150,000 (11,000)
	Installed capacity (1000 m^2)
Botswana	50
Malawi	4.8
Mauritius	40
Namibia	24
Seychelles	2.4
South Africa	500
Zimbabwe	10

Source: (Adapted from Karekazi and Kithyoma 2003)

Table 7.3 Large hydropower projects as a percentage of installed capacity in selected African countries

Country	Installed capacity (MW)	Installed capacity (%)
Mozambique	2075	100
Uganda	260	98
Zambia	1786	93
Malawi	242	90
Ethiopia	424	88
Kenya	885	70
Namibia	387	62
Tanzania	655	58
Zimbabwe	1961	33
Mauritius	425	12
South Africa	38,517	0.01

Source: (Adapted from AFREPREN/FWD 2002)

estimates that Africa has sufficient hydro capacity to provide for its needs, with an adequate allowance for export (IRENA 2013). However, despite the significant amount of hydropower potential, only 7% of the continent's hydro resources have been harnessed. Table 7.3 shows large hydropower facilities as a percentage of installed capacity in selected African countries.

There are numerous hydropower projects being planned, mainly in SSA. These projects range from small, mini, and micro projects to large-scale hydropower plants. The utilisation of small hydropower plants is mostly found in rural areas, particularly in remote and isolated regions because it is a cost-effective, off-grid solution for electricity generation. These plants are driven by small water turbines that are powered by water diversions from a rudimentary dam. According to IRENA (2015b), the average cost of installation for small-scale hydropower in Africa is US $3800/kW. Large hydropower projects include the Tekezé Dam, Ethiopia, with 300 MW installed capacity and the hydropower stations in Mozambique on the Zambezi River.

Two large hydropower projects are under construction: The Grand Inga project and the Great Millennium Renaissance Dam. The Grand Inga project is located on the Congo River and will be the biggest hydro facility in the world. It is expected to have a total installed hydropower capacity of 40 GW and is being developed in eight phases. Inga 3 is the current development phase of the project with a potential electricity generation capacity of 7.8 GW, out of which 4.8 GW is underway towards commissioning in 2030 (AEEP 2016). In Ethiopia, the Great Millennium Renaissance Dam is located on the Nile River with a projected 6 GW of hydro generating capacity. At present, the total hydropower capacity in Africa is 525 MW, with 209 MW in eastern Africa and other countries having individual capacities of less than 10 MW (IRENA 2015a). Figure 7.2 shows the hydropower generation capacity for various African countries and the percentage of generation potential to the technical potential.

Wind Energy Technologies

Africa has one of the best potentials for wind power production, because of its large coastline and island states (Tiyou 2016). The energy from wind is principally harnessed by wind turbines from the aerodynamic force of high-velocity winds and this energy comes in different scales. The technological approach and applications in Africa are typically not complex and require low maintenance, provided the annual average wind speed is greater than 4 m/s for small wind turbines and 6 m/s for utility-scale wind power plants (Albæk 2015). Water storage capacity is usually

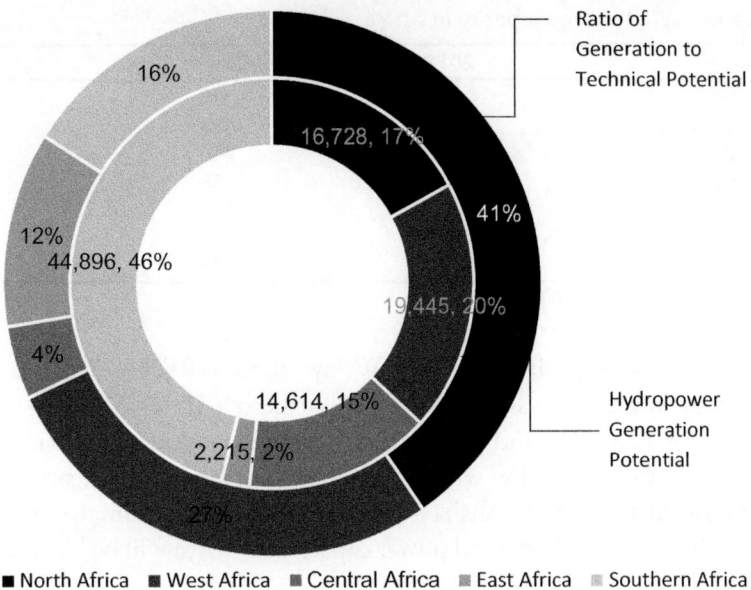

Fig. 7.2 Hydropower generation potential (GWh) and ratio of generation to the technically feasible hydropower potential (%). (Data from Hydropower and Dams 2014)

integrated in these technologies to ensure a continuous supply of water. Southern Africa has over 300,000 units of such wind turbines in operation (IRENA 2015a). The power can be used for on-grid or off-grid power generation or applications, such as water pumping for agricultural purposes.

The total installed capacity for wind power generation by the end of 2011 in Africa was 998 MW, and by the end of 2016, a new capacity of 2728 MW was installed, making a total of 3726 MW. The countries in which they are located are shown in Table 7.4. In 2014, Morocco had the largest installed wind capacity in Africa, followed by South Africa, which had a significant increase from the previous years. In the eastern part of Africa, a 300 MW Turkana project is underway. Also, a total of 140 wind farm projects, with a capacity of 21 GW, are under construction and expected to come into full operation by 2020 (GlobalData 2015). In Egypt and Morocco, wind power capacities of 7 GW and 2 GW are expected to be installed, respectively. South Africa also plans to achieve a

Table 7.4 Wind energy capacity in Africa

Country	2011	End of 2016
Morocco	255	798
South Africa	10	1473
Egypt	550	750
Tunisia	53	245
Ethiopia	81	324
Others	49	136
Total	**998**	**3726**

Data Source: (IRENA 2017)

total installed capacity of 8.4 GW by 2030 (IRENA 2015a). The Ashegoda wind farm situated in the northern part of Ethiopia has a capacity of 120 MW, and the country is already working on the installation of additional 51 MW wind farms located in two different sites in the southern part of Addis Ababa (Waruru 2014). It is estimated that by 2030, the total installed wind power capacity in Africa will be between 75 and 86 GW.

Geothermal Energy Technologies

Geothermal energy has the advantage of minimal to negligible emissions, especially for closed systems and processes where water is reinjected back to the earth's crust. According to Karekezi and Kithyoma (2003), geothermal plants require 11% of the total land use that is required by coal-fired plants, and 12–30% of land space used by other renewable technologies. As such, geothermal plants have the advantage of reduced land requirement as this energy can be derived at high and shallow depths, provided there is sufficient concentration of steam at high pressure. The recovered steam, typically from drilling wells, is then used to drive turbines for electricity generation. Africa is estimated to have 9000 MW of geothermal energy; however, only about 59 MW has been tapped, mostly in Kenya with 57 MW, and the remaining 2 MW in Ethiopia (Karekezi and Kithyoma 2003).

Kenya has been exploring geothermal energy since 1956 and there has been increasing utilisation. A geothermal plant was installed as part of the Olkaria project to explore nearly 103 wells of various depths of

180–2600 m (Karekezi and Kithyoma 2003), with plans to increase the capacities to 576 MW. Going forward, Tanzania and Ethiopia aim to increase capacities to 640 MW by the end of 2018, while Djibouti aims to harness geothermal energy by 2020 (IRENA 2015a). These initiatives are supported by international development programmes such as the Geothermal East Africa Initiative, a project funded by the German Development Bank to support the exploration of geothermal energy. Other programmes are supported by UNEP and World Bank, in particular for the Rift Valley regions (IRENA 2012).

Bioenergy

In Africa, biomass is utilised for various applications, including direct combustion, gasification, cogeneration, biogas production, ethanol production, and briquetting, on small and large scales. Agro-based industries, including sugar, paper and pulp, rice, and wood industries, use cogeneration (heat and power) to meet their energy needs. For instance, there is a high potential for electricity generation from bagasse (the dry residue from sugarcane after juice extraction) since the equipment for cogeneration can be easily integrated into the design of sugar factories. Countries that produce and export sugar, such as Malawi, Ethiopia, Madagascar, Zimbabwe, and Swaziland, can therefore harness energy from biomass. It is estimated that about 16 countries in SSA can use bagasse-based cogeneration to meet current electricity requirement (Deepchand 2002; Karekezi and Kimani 2002). Mauritius is a typical example, where cogeneration meets over 20% of the country's electricity requirements. Their sugar industry is self-reliant in the generation of electricity from the extensive use of cogeneration, and this enables them to sell excess power to the national grid (Karekezi and Kithyoma 2003).

Over the years there has been development in more efficient ways to utilise biomass for more reliable bioenergy supply. In Ghana, Ethiopia, Congo, Namibia, Swaziland, and Tanzania, 11 wood-based power plants have been installed and are in operation, having a total capacity of 30 MW, and a few more are under construction (Platts McGraw Hill Financial 2015). In the rural parts of Africa, most cooking, heating, and lighting needs are met using cookstoves based on wood fuel. Charcoal is used as a household fuel, mainly in poor urban areas, where it is preferred to wood fuel, due to its easy storage, lower level of smoke emission, and high-energy content. Thus, considerable efforts are being made towards the improvement of small-scale biomass systems along with the implementation of cookstoves and charcoal kilns that are reliable and less harmful to the environment for rural households and poor urban regions in SSA. Kenya and Somalia have well-founded and reliable cookstove programmes, and the trend appears to be increasing currently to other parts of Africa. The use of efficient cookstoves in Africa has also been on the increase with 36% usage in Kenya and 50% usage in Rwanda, and a total of 2 million efficient stoves have been deployed in Africa (Global Alliance for Clean Cookstoves 2014).

Biogas is another source of energy that has drawn attention over the years. Biogas is produced via the anaerobic digestion of organic wastes such as sewage, animal waste, and food processing wastes, which are readily available in many Sub-Saharan rural areas (Karekezi and Kithyoma 2003). The feasibility of this technology has been proved by tests conducted and pilot projects, but there are some problems, especially the collection of dung for use in a large-scale production. Also, farmers who keep livestock on a small scale found it difficult to gather enough feedstock for their bio-digester unit so as to ensure steady power generation for lighting and cooking. And lastly, the cost of the smallest biogas unit is not affordable for the residents of rural poor African countries. These factors have limited the growth of this technology, although there is a consensus that larger combined septic tanks and biogas units that are used by bigger institutions, such as schools and hospitals are more feasible than small-scale biogas digesters (Karekezi and Kithyoma 2003).

The new trend for cooking fuels in Africa is towards ethanol and ethanol gels (IRENA 2015a). Efficient ethanol stoves are being prepared for distribution in Mozambique and about 1 million litres of gel fuel is generated in South Africa annually. About 60% of the plant for the ethanol programme in Zimbabwe was locally produced, and the plant has been in operation for 20 years with little maintenance required (Karekezi and Ranja 1997). In Malawi, Zimbabwe, and Kenya, a blend of ethanol and gasoline for use in motor vehicles has been generated from ethanol programmes. The production of gasoline and ethanol has made a significant contribution to the country's economy. The ethanol programme in Zimbabwe had a capacity of producing about 40 million litres, and measures were taken to increase the yearly output to 50 million litres (Karekezi and Kithyoma 2003). In Kenya, the total investment cost for the ethanol plant is approximately US $15 million. The optimum production from this plant is about 45,000 L/day (Karekezi and Kithyoma 2003). The plant uses molasses, that would have been disposed of as waste into nearby rivers, to produce ethanol. The ethanol is mixed with gasoline at a ratio of 1:9 and sold locally. The ethanol programme in Kenya has been experiencing a considerable loss yearly since it has been commissioned, because of the influence of the government on the low cost of the retail price; insufficient and inconsistent plant maintenance and operation; opposition from local subsidiaries of multinational oil companies; and an adverse high currency exchange rate that has increased the local cost of servicing the loan that funded the plant. However, the plant has generated approximately 1000 rural jobs. For successful expansion of renewable energy programmes, there is therefore the need to minimise any barriers, such as financial, legal, and technological constraints.

The next section further expands on some applications of clean energy technologies using case studies.

Main Applications of Clean Energy Technologies in Africa

Cooking

According to the World Health Organization (WHO), there are over 2.5 billion people in developing countries who depend primarily on biomass fuels (e.g. fuelwood, charcoal, and agricultural waste) for cooking (IEA 2006). A large proportion of this population are resident in rural communities, particularly in India, China, and SSA. Although, the use of biomass fuels is largely encouraged, the unsustainable consumption of the fuels in inefficient cookstoves is a global concern, because of the risk associated to human health, land degradation, and environment pollution. Traditional cooking is still most widely used and practised in rural settlements in many parts of Africa, by placing a pot on stones as support then placing the charcoal underneath as the source of fuel. This cooking method is not environmentally benign and subjects its users to health risks from inhalation of smoke and particulate matter. Thus, the use of fuel wood and charcoal in exposed, inefficient stoves with emissions of black carbon and toxic gases is largely associated with indoor pollution and millions of premature deaths recorded annually, especially among women and children under the age of five. A report by the WHO (Rehfuess 2006) estimates 1.5 million annual premature deaths are due to the use of solid fuels for cooking, numbers that are close or comparable to those caused by malaria, tuberculosis, or HIV/AIDS. To address these problems, the development and improvement of clean stoves and energy conversion technologies for cooking are underway in many parts of the world for such communities.

Two main clean energy technology approaches in cooking involve the development of efficient, clean stoves or devices, and the sustainable production and use of clean cooking fuels. Typical efforts to produce clean fuels that can burn efficiently, even in conventional cooking stoves, include the production of liquid or gaseous biofuels from locally available renewable energy source and exploitation of solar energy for cooking. For instance, the production of ethanol is largely encouraged from non-food crops (e.g. algae, jatropha, wood, grasses, agricultural residues, crop products, and waste materials of food crops), although the use of food crops (e.g. sugarcane, sugar beet, sweet sorghum, cassava, and potatoes) is still widely practised. Ethanol gel is produced by mixing ethanol with cellulose to generate a clean-burning fuel that is not susceptible to spillage, and it is safer than liquid ethanol and kerosene. Ethanol gel stoves have been developed using a single or double burner system; the heat level can be controlled and they have a silent operation. Their use has been observed to be competitive against the use of charcoal and wood in Ethiopia, Mozambique, Malawi, South Africa, and Senegal in 2004 (Utria 2004). However, cooking with ethanol gel is more expensive than cooking with liquid ethanol (IRENA 2015a). The average cost of ethanol gel is approximately three times the cost of liquid ethanol, particularly blended liquid ethanol. Other cooking fuel options include the use of liquefied and pure gaseous fuels. Liquified Petroleum Gas (LPG) is increasingly used, particularly in urban areas. These fuels are provided in

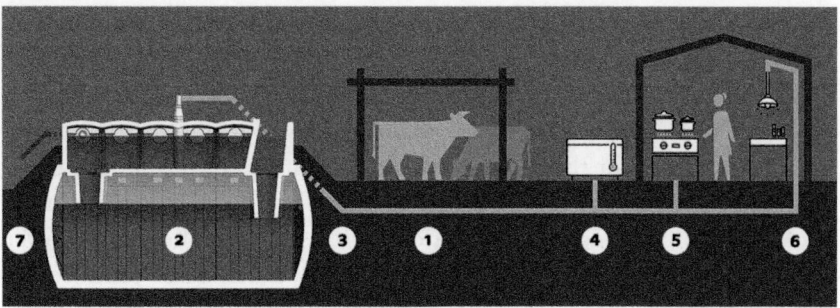

Fig. 7.3 Flow diagram of the SimGas Biogas System. (1) Manure from livestock, (2) anaerobic digestion, (3) piping, (4) milk chiller, (5) cookstove, (6) biogas lamp, (7) organic fertiliser. (Source: © SimGas 2018)

conventional low-pressure gas burners with cylindrical canisters (Global Alliance for Clean Cookstoves 2017).

There are several ongoing biogas projects on a small and community scale, such as the SimGas Biogas System Case Study as shown in Fig. 7.3. These systems are designed to produce biogas via anaerobic processes.

Case Study 1: SimGas (Simgas 2017)

The SimGas Biogas System is designed for rural farm households in developing countries. The unit consists of a digester (2) that converts manure (1) to biogas (3) via anaerobic processes and receives slurry manure from farm animals. The resulting biogas is then piped into the user's milk chiller (4), cookstove (5), and gas lamps (6) for cooling, cooking, and lighting respectively. The residue is fully digested and can be applied to the farm as an organic fertiliser (7). The system is easy to install and has a modular design, that is inlet, expansion chamber, gas holder that enables the farmer to downscale or upscale.

The effort to design efficient, less polluting, clean stoves or devices is aimed at reducing indoor pollution and other associated environmental hazards. This includes the design of one or more of the following structural components such as: (i) a chimney system that improves the movement and transportation of the exhaust gas from the cooking stove to the outside of the home. This does not necessarily reduce environmental pollution to the surrounding area; however, it does limit indoor pollution and consequent adverse health impacts; (ii) air fan that is powered by a thermoelectric device, battery, or mains to improve circulation and provide sufficient air for the fuel to completely combust; (iii) modern cooking system with improved, efficient burn, such as a gasifier, solar cookstoves, and insulated combustion systems. The gasifier stoves, often with an updraft system, force the incomplete burnt gases produced to pass through the flame and this improves air and gas mixtures while reducing emissions from the stoves. The insulated combustor reaches high temperatures and uses high thermal conducting materials. This reduces emissions because of improved heat transfer processes and limited heat losses. The solar cookstove takes advantage of available solar energy, particularly in countries with a hot climate, to power cooking devices. These systems have zero emissions and no fuel costs, but are limited because of the associated capital costs and constraints of constant solar energy supply. Modern electric cooking systems are also used in urban cities but they are limited because of low energy access in developing countries.

Case Study 2: African Clean Energy (ACE) 1

This is a solar biomass cookstove that claims to burn any type of biomass fuel efficiently for cooking using excess power supply (African Clean Energy 2017). It uses the processes of gasification to convert the biomass fuel to hot clean gas and this is aided by an air fan. The fan blows air into the burning chamber via small holes at the bottom and the top of the unit. This forces the fuel to burn until a high temperature of ~1000 °C is reached. Clean gas is produced when the hot flue gas at the burning chamber mixes with air at the top of the unit. The air fan is powered by a battery power supply that can last for up to 20 hours of cooking. Energy is supplied to the battery power supply via a small 9 V 5 W solar panel and this can be used for charging a mobile phone or for lighting via a light-emitting diode (LED) port. The unit burns 15 g/min wood pellets at full fan power and 7.5 g/min at low fan power. It has been deployed in nine countries including Kenya, Lesotho, South Africa, Uganda, and Zambia. A 70 L pot capacity ACE 1 unit currently sells for US $150 plus shipping costs of US $70.

Case Study 3: InStoves

The InStoves is considered to be an institutional-scale unit and has a capacity in the range of 20–100 L. It can be used for sterilising medical supplies, pressure cooking, preparing safe drinking water, and so on. The combustion chamber is insulated with alloy metal and can reach temperatures of 1100 °C, thereby burning off all volatiles and products of incomplete combustion. Compared to conventional stoves, it reduces biomass fuel use by 75–90% and reduces the production of toxic emissions by 90% or more. It also has an integrated chimney unit that pulls all gaseous emissions outside of the immediate environment. The unit is mobile and considered durable for 5 years (no maintenance) or longer for replacing parts (Instove 2017). The unit costs between US $475 and US $1095 for 20–100 L capacity. It has been deployed in several African countries including Kenya, Nigeria, Zambia, Uganda, Zambia, and Sudan. The unit is manufactured in Nigeria with some parts supplied from the United States via a "Stove-Factory-in-a-Box" mass production model.

Agriculture

Agriculture is one of the forces driving economic growth and development of African countries, yet the continent is unable to meet its own need, because of a lack of infrastructure and technology. The region battles the challenges of climate change, diminishing natural resources and rapid environmental degradation, and as such food security is a serious threat. This is further compounded by rapid population growth, food prices, and consumption patterns. Since two-thirds of the populace depend on agriculture for subsistence (Garcia et al. 2006) and the continent expends billions of dollars on importation of food (about US $35 billion in 2011) (Africa Progress Panel 2014), the development and adoption of innovative, sustainable, and integrated clean energy technologies is one of the intended approaches of boosting food security, and improving the yield of farm products and productivity of farmers. These include integrated energy systems for pest management, water use, wastewater treatment, fertiliser application, pollution prevention and/or control, energy management, precision farming, data processing, and so on.

Some of the clean energy technologies developed as part of "Powering Agriculture", an initiative that aims to address the problems of energy poverty and food insecurity in low-income countries, include:

(i) Biogas-powered systems (e.g. SimGas and EvaKuula) for preserving farm products. These systems utilise waste such as animal slurry, slaughter, and agricultural wastes to generate biogas. For instance, the SimGas uses the biogas generated from dairy farm animals to power a chiller that ensures milk longevity, thus improving the quantity and quality of milk produced by domestic-scale farmers. Four prototypes were tested in Tanzania. EvaKuula pre-treats milk mildly and further provides cooling by evaporation. This reduces milk spoilage and ensures that milk produced by local farmers stays fresh overnight. So far, two units have been deployed in Uganda, but research and development (R&D) is ongoing to further develop the product. The biogas savings can save income that would otherwise be spent on kerosene, wood, and charcoal, while enhancing milk production. A similar technology developed by the Horn of Africa Regional Environment Centre and Network (HoA-REC&N) uses biogas generated from coffee pulp and husk in a bio-digester to reduce drying time via an infrared dryer technology. Four of these units have been tested and deployed in Ethiopia.

(ii) Solar-powered pumps and irrigation systems (e.g. SunCulture AgroSolar Irrigation Kit, Sunflower Pump, and MoneyMaker) that aid agricultural productivity of smallholder farmers. These systems employ solar power via PV panels to power small-scale, easy-to-fit and install irrigation pumps for diverse agricultural farming activities. Irrigation, as opposed to rain-fed systems, ensures that the farmer can have considerable produce all year round; thus, increasing productivity and income for farmers. The units are also offered on a flexible platform to minimise the risk associated with upfront capital costs. For instance, the MoneyMaker Pump integrates a "pay as you go" unit that offers a flexible payment method. The Sunflower Pump, a system considered as easy-to-maintain because of the simple piston pump arrangement, is offered to farmers on finance. Thus far, the Sunflower Pump Company has produced more than 400 units for Kenyan farmers, MoneyMaker has sold more than 300,000 units across Africa, and the SunCulture AgroSolar Irrigation Kit has provided irrigation for 187 farmers, as well as trained 25 technicians in Kenya.

(iii) Thermal plants to produce energy for various agricultural activities include fruits, oil, cereal, and wood processing; dairy pasteurisation; water purification; irrigation and so on. The ThermoBoo thermally treats bamboo wood against rots, insects, and for further wood processes. This chemical-free process plant self-generates heat and electricity from bamboo dust via combustion processes. The Village Industrial Power (VIP) Steam Plant processes agricultural waste to produce power for small villages. So far, five field tests have been carried out in Kenya, Benin, and Tanzania, which produced 7.5 kW of electricity and 40 kW of thermal energy for agro-processing facilities.

(iv) Solar refrigeration solutions preserve post-harvest produce and reduce food spoilage. For instance, the SunChill™ removes field heat from crops at about 50 °C following harvest to provide cooling of about 10 °C. The unit was pilot tested in Mozambique but is currently undergoing further development into a commercial product. SunDanzer employs solar energy via PV cells to provide cooling in chest refrigerators and for small-scale, cold-chain applications. This technology has been installed for 38 Kenya dairy farmers whose products feed into two cooperatives.

Electricity

Modern electricity is a scarce resource in SSA, particularly for rural communities. The continent is estimated to have over 600 million residents living without electricity (IEA 2014), about two-thirds of the African population and more than 40% of the people in developing countries. To improve energy access, there are concerted efforts to provide clean, affordable, and sustainable energy products across multiple organisations, platforms, and initiatives. These products are typically pilot, innovative technologies (developed within or outside of Africa) that would be upscaled. They include large-scale, grid-based electrification and small-scale (off-grid, mini-grid, and stand-alone) electrification systems. The grid-based systems have the advantage of economies of scale and can significantly increase the national electricity consumption per capita of any country. However, they require extensive expansion of generation, transmission, and distribution capacities, infrastructures that are currently lacking in many developing countries, even for highly populated urban cities. Therefore, there is currently a strong focus on providing small-scale, domestic solutions for low-income, rural communities with a desperate need for electricity. This includes the provision of low-power rating, low-cost, high efficiency, off-grid appliances for lighting, mobile phone charging, cooking, heating, cooling, entertainment, and so on.

Lighting Africa, an initiative and contribution of the World Bank Group to the Africa Renewable Energy and Access (AFREA) programme, Sustainable Energy for All (SEforAll), and Global Lighting and Energy Access Partnership, has sold almost 15 million solar lighting products in Africa since 2009, yet the market is not saturated and continues to experience growth (Lighting Africa 2018). These off-grid lighting products are solar-powered devices, and typically based on pico-photovoltaic systems and LEDs. They include components such as PV panels for solar energy collection and batteries and/or inverters to utilise the power. The products vary from small, portable solar lanterns for small households, to large systems that serve community-scale needs. To improve sustainability, the products are usually supported with financing facilities, R&D funding, quality assurance and standardisation, market intelligence, consumer awareness and training, as well as support for policy development. Another example is the Solar Lantern Library in Senegal, which is equipped with about 5000 solar lanterns to provide lighting for 6000 students (SolarAid 2014). The units are rented from the solar library and alternative purchase options are also provided.

There is large-scale, grid-based electrification across many parts of Africa, as well as mini grids for rural and peri-urban areas. Many of these projects have been commissioned in the last decade and are undergoing installation. The Ouarzazate Solar Power Station in Morocco (Noor Complex), which is regarded as the largest CSP complex in Africa, with a capacity of 160 MW, started operation in 2016. Additional 200 and 150 MW CSP facilities, respectively known as Noor II and III, are expected to be in operation in 2018 (NREL 2017). Two similar projects in South Africa are currently being developed with respective capacities of 50 and 100 MW (Solarpaces 2017). The Ingula Hydroelectric Plant in South Africa is expected to produce 1100 MW, once in operation.

Case Study 4: Solar Sister Lantern and Cell Phone Chargers

The solar sister lantern and cell phone chargers are distributed to women across an established network in Africa using a "micro-consignment" model. Here, each woman receives training, marketing materials, and inventory of product to sell directly to the consumer. Each solar lantern is said to cost US $18 and saves ~US $160 of fuel over a 5-year period. The solar lantern with mobile phone charger costs US $45 and saves ~US $225 of fuel over the same period. Both products have the additional benefits of increasing the working hours of consumers and providing a sustainable income. They reduce indoor air pollution and loss of reading time for children. Based on the social business model, Solar Sister has empowered more than 170 entrepreneurs across Africa and provided power to 30,000 Africans or more (www.solarsister.org).

Sanitation

The lack of access to modern sanitation facilities is of growing public health concern worldwide. About 2.4 billion people have no access to improved sanitation and nearly 1 billion people dispose faeces directly into the environment (WHO/UNICEF 2017). In developing countries, it is estimated that 90% of the excreta generated by humans end up in watercourses and rivers, due to inappropriate dumping and open defecation of faeces into the water bodies (Danso-Boateng et al. 2013). In SSA, the number of people without access to improved sanitation total about 695 million and more than half of these are in the rural areas. Even where modern sanitation systems exist in urban areas, there are reports of land degradation and groundwater contamination because of wastewater leaks from improper installation and maintenance of septic systems (Kayembe et al. 2018; Laffite et al. 2016). As such, poor sanitation is considered to be one of the leading causes of transmission of waterborne pathogens and epidemic breakouts of waterborne diseases (e.g. dysentery and cholera). It is also well associated with the spread of parasitic worms worldwide. Based on a report by Liu et al. (2000), it is estimated that more than 2000 children under the age of five die of diarrhoeal diseases daily and more than 1.5 billion people are infected with parasitic worms worldwide (Nelson et al. 2015). Poor sanitation is thus an impediment to development and requires the advances of modern sanitary systems.

There are ongoing efforts to develop novel sanitary systems that can more efficiently and safely treat human excreta than the conventional toilet systems, for example, Loowatt Toilet, Nano Membrane Toilet (Onabanjo et al. 2016a), and Eco-San Toilet. This is because the conventional systems with centralised design approach are cost-, resource-, and energy intensive. In other words, they require extended sewer infrastructures or connections to waste water treatment plants, significant amounts of water resources for flushing and transportation, and supplementary power to facilitate treatment and disposal. The novel sanitary systems are expected to integrate clean energy systems that can convert the excreta to useful by-products such as bio-oils, char, water, fertiliser, heat, and/or electricity (Onabanjo et al. 2016b). Some of the main clean energy approaches include processes such as gasification, hydrothermal liquefaction, pyrolysis, combustion, and anaerobic digestion. These systems are intended for toilet designs and approaches such as urine diversion systems, urinal-focused facilities, or mixed collection systems with waterless toilet that are still under development.

Apart from domestic solutions, there are ongoing efforts to develop large-scale, integrated community sanitary systems or solutions. For example, in the Nano Membrane Toilet, human excreta from individual collection systems would be transported or transferred to a retrofitted or tailored (sewage) biomass plant for thermochemical conversion to ash, electricity, and/or heat (Onabanjo et al. 2016a). These solutions are of importance especially in rural areas, since many villages and communities are wide distances apart. Here, a centralised waste treatment plant would not be an ideal solution. These technological approaches can prevent the open dumping of faecal material in the environment and the recovery of those materials for energy production, which can facilitate rural electrification. They can also resolve many other health and social imbalances.

Challenges, Barriers and Approaches to the Development, Transfer and Diffusion of New and Innovative Clean Energy Technologies

Meeting the needs of a changing society is a challenge for manufacturers, as the development of an innovation into usable products is marred by many failed attempts (McCrum 2015). Clean energy technologies are not an exception, particularly for developing economies where transfer and diffusion of a new technological innovation is more subject to the ability of individuals in the society to adapt to change. The transfer and diffusion of innovations involve the fundamentals of the development of jointly constructed understandings of the world, the gradual integration of a technology, and the social process of dissemination among consumers (Davis 1989; Samaradiwakara and Gunawardena 2014; Zhang et al. 2015). Diffusion of Innovation (DOI) theory gives account of the movement of an innovation from invention to adoption or rejection (Dillon and Morris 1996). This is expressed in terms such as transfer, adoption, acceptance, spread, use of innovation (WHO 2010). Technology transfer is the process of movement of developed technologies, expertise, and basic inventions into useful and commercially relevant products between organisations. The cycle includes not only the commercialisation of products but also the delivery of knowledge, expertise, and techniques. The process of dissemination is gradual between consumers and users, and involves the creation of awareness and knowledge about the idea or innovation (Robinson 2009; Roman 2004). Adoption, on the other hand, is influenced by a person's awareness of the need for a new idea, trial of the innovation, and finally the decision to accept or reject the technology (LaMorte 2016). As such, technology adoption rates within concerned groups differ, while some are early adopters, others may adopt at the later stage (Dibra 2015; Stieninger and Nedbal 2014).

According to Rogers Theory, innovation is attributed to five different user-perceived behaviours: relative advantage, compatibility, complexity, observability, and trialability (Dibra 2015). Relative advantage refers to the extent to which a technology can offer benefit, compared to the initial technology (Lee et al. 2011); compatibility is the extent to which innovation fulfils the needs of the user; complexity is the perceived level of difficulty in understanding technology, i.e. the ease of use of technology by the users and is an important factor for consumer adoption of innovation (Dearing 2009; De Civita and Dasgupta 2007); observability is the degree to which adopters perceive the benefits of an innovation as being noticeable; and trialability is the partial use of an innovation with a negligible degree of obligation. When an innovation is under trial, adopters may not be totally committed. All the above-mentioned user-perceived behaviours can influence the transfer and diffusion of any technological information, as such societal perception cannot be overlooked.

Barriers to Diffusion of New and Innovative Clean Energy Technologies

Challenges to the spread of new technologies are generally similar but for some differences in characteristics from region to region. For example, barriers to diffusion of new technology in developing countries are different from those of the developed ones. Some of the barriers affecting developing nations are believed to be associated with weak governance and a lack of concrete policies and projection (Kirkland and Sutch 2009; Painuly and Fenhann 2002; WHO 2010). Generally, the barriers to diffusion of new and innovative clean energy technologies can be broadly classified into financial, institutional, environmental, technological, and information barriers (Painuly and Fenhann 2002). Policy factors include guidelines on import and export, government funds, promotions and incentives, market regulations, taxes, and trade laws. Financial barriers can result from unavailability of funds, lack of natural markets and high cost of foreign technology. Others can include inadequate knowledge of foreign markets, weak intellectual property, poor after-sales service, and so on. These barriers are discussed in the subsequent subsections.

Financial Barriers

Clean technologies exhibit higher uncertainty, regulatory dependency, and require huge investment capital for start-up compared to other energy technology alternatives. This makes them unattractive for investors (Polzin 2017). Moreover, most governments in developing economies have poor incentives, such as tax reduction and subsidies for clean technology, which could hinder the use and spread of the technology, especially in Africa. The use of clean energy technologies can be improved if encouraged for domestic use in the form of a stand-alone technology. However, access to micro financing and credit facilities are highly restricted and sometimes not available at all; when they are available, the funds have high interest rates (Nnaemeka and Nebedum 2016). Most clean energy technologies, such as wind and solar, are intermittent in nature with unpredictable output (Ayodele and Ogunjuyigbe 2015). Furthermore, there are concerns among investors and utility operators that clean energy technologies with variable output may increase the operating costs, i.e. cost of providing backup for the intermittency of clean technology during times of low solar irradiation or wind speed (Ayodele et al. 2012). It is often believed that this does not make investment sense hence resulting in reduced financial commitments.

Institutional Barriers

Government policies and regulations play a significant role in the diffusion of innovations. Government regulations and conditions imposed on manufacturers in the form of environmental tax, energy tax, tariffs, and levies all have an effect on the supply and demand of clean energy technologies (Painuly and Fenhann 2002; Poole and Buckley 2006). Also, the lack of information sharing among the utility operators on the experience gained in handling the intermittent nature of renewable energy is hindering the progress and acceptance of clean energy technologies. Most responsible agencies are unaware of modern trends and recent breakthroughs in clean energy technologies. In addition, stakeholders in clean energy technologies in many parts of Africa have a slow response to new trends in technology as a result of poor attitude to capacity building. All these issues have slowed down the adoption and diffusion of clean energy technologies (Hornor and van Gerven 2015; Ikejemba et al. 2017; Wojuola and Alant 2017).

Environmental Barriers

The business environment, which includes community culture, values, beliefs, and norms, can inhibit or facilitate the diffusion of innovations such as the clean energy technologies. Barriers, such as physical hindrances in space, illumination, and limited space to display ideas, can pose barriers to technology spread (UCSUSA 2017). The lack of commercial innovation by the clean energy technologies vendor in using factors relevant to the community, such as their culture, belief, and faith, to converse with and persuade customers to adopt their products, inhibits the spread of clean energy technologies (Liyanage et al. 2016).

Technological Barriers

Technological barriers are influenced by restricted access to R&D (Bozeman et al. 2008). Adequate data are required by policy makers, investors, governments, and non-governmental organisations (NGOs) to take informed decisions on optimal investment in clean energy technologies and to understand their possible impact when integrated into the grid. These data and information are lacking in Africa and this could be hindrance to appropriate policy formulation to incorporate clean energy technologies into resource planning. Another technical issue is the plug and play technology needed to interface clean energy technologies, such as wind turbines, solar PV, and small hydro generators, into the grid. Despite the giant strides recorded in clean energy technologies in recent years, development of a standardised generic model for clean energy technologies and guidelines that are universally acceptable are still not available. The plug and play technologies that can enhance grid integration of renewable energies are still underdeveloped and the levelised costs of energy, especially for small-scale stand-alone clean energy technologies are still very high compared to the alternatives. This has made the acceptance and spread of clean energy technologies in Africa a slow process (Ayodele and Ogunjuyigbe 2015). Diffusion of innovations becomes easier through local capacity building and exhibition of new technologies (Nisbet and Collins 1978). Many manufacturers lack insight into the local demands of economies in Africa. Therefore, understanding and meeting the energy requirements of the developing countries with the required technology becomes a challenge. Local capacity building through the transfer of knowledge of clean energy technologies is lacking. This inhibits the diffusion of technologies because it is generally believed that it is not a good idea to subscribe to any technology without local experts or understanding. Any technical problems without a local expert will render a product useless as it is difficult and expensive to have access to foreign experts. Another technological barrier to clean energy technologies is the complexity of the technology, which may hinder its suitability for use by many consumers (Mazurkiewicz and Poteralska 2017).

Information Barrier

Inadequacy of information to users of clean energy technologies may inhibit their ability to diffuse and spread (Liyanage et al. 2016) especially in African communities. The sustainability of any technology rests on clarity of purpose to the entire populace. Clarity is defeated when the basics, purpose, and process-related elements of the technology are not well shared, understood, and communicated (Mazurkiewicz and Poteralska 2017; Nisbet and Collins 1978). Most of the clean energy technologies provide little or no information about their operations in relation to the conventional technology, hence little is known about the advantages of the technologies in comparison to the conventional ones. For instance, inadequate information about the hybridisation of renewable energy sources such as solar and wind technologies may make some consumers see these sources as being ineffective and not useful when the weather is not windy or sunny, whereas, a combination of these technologies is consistent.

Mechanism for Promoting the Development, Transfer and Diffusion of New and Innovative Clean Energy Technologies

In order to promote the development of clean energy technologies as well as aiding the diffusion of technologies in Africa, there is a need to institute mechanisms that will accelerate the growth and transfer of technology. This is necessary and presently the case as the private sector investment in R&D is in its infancy in Africa (UNECA 2017; UNIDO 2017). As part of the mechanism for technology development, governments need to foster collaboration among academia, industry, and government itself. This can be achieved through the establishment of sustainable policies and a legal framework that will boost collaboration among these stakeholders. These mechanisms may include collaborative funding between academia, industry, and government, firm-friendly regulations that can boost clean energy technologies development, establishment of a legal regulatory structure governing intellectual property rights that will strike a balance between manufacturers' and users' interests, and an exchange of knowledge and best practices (Flamos 2010; Osok 2010; UNECA 2017).

Mechanisms towards the development and transfer of clean energy technologies can be broadly categorised into institutional arrangements and support mechanisms for the development and transfer of clean energy technologies.

Institutional Arrangements

As part of the mechanism for clean technology development and transfer, governments need to establish a consultative board drawn from various government agencies for devising viewpoints and perspectives on technological growth, spread, and dissemination. This will enable the coordination of activities across different agencies. Also, there is a need for the creation of a coordinating committee tasked with the development, diffusion, and use of clean energy. The coordinating committee could be contributory to inspecting organisations, providing technology dissemination, undertakings, and managing technology-related costs. The committee's duties and responsibilities could also include regional forums for promoting best practices with support from independent experts. Networks and partners involved in the promotion of technology development and transfer in Africa include among others: African Technology Development and Transfer Network of UNECA, African Agriculture Technology Foundation (AATF), Alliance for a Green Revolution in Africa, African Renewable Energy Alliance, and the African Network for Drug and Diagnostic Innovation (ANDI). There is also a need for instituting network centres and hubs for clean energy technology such as NEPAD Biosciences Hubs, the ANDI Network of Centres of Excellence, members of the African Technology Development and Transfer of UNECA, and the Pan African University Centres of the African Union. These centres could be established regionally to evaluate the level of each country's participation in clean energy technologies development. The purpose of the Centres of Excellence would be to stimulate collaborative R&D and training and to influence the exchange of resources for the development, diffusion, and transfer of clean energy technologies between the government institutions and the industries.

As part of the mechanism for the development of clean energy technologies, the United Nations established the United Nations Industrial Development Organization (UNIDO) and tasked it with the responsibility of stimulating socially inclusive and environmentally friendly industrial development, with a well-established ground for expertise in promoting a lucrative and profitable investment, and transfer of knowledge and technologies. This includes the constitution of National Cleaner Production Centres situated in various developing countries for linkages through a Resource Efficiency and Cleaner Production network to support government industries, especially small-scale enterprises, to implement viable production techniques. Other specialised networks established by UNIDO include International Technology

Centres in various countries for the development of specific clean energy technologies (UNIDO 2017). African leaders need to be more proactive to ensure that the desired result of these specialised agencies is realised for the development and diffusion of clean technologies.

Support Mechanisms

One of the support mechanisms that could support the rapid growth and spread of clean energy technologies is adequate funding. A reasonable funding mechanism that could back the reduction in costs associated with clean energy technologies development and its diffusion should be put in place. Manufacturers in developing countries need to have access to adequate fund with the support of governments, development partners, international organisations, the business sector, and civil society (UNECA 2017). Responsibilities to be funded include licence fees and training costs related to clean technology transfer levied by technology owners. However, specialised agencies such as UNECA and the African Development Bank could also render financial support to developing countries for the development and diffusion of clean energy technologies.

Other means of supporting the growth and diffusion of clean energy technologies include capacity building, such as individual, private, public, and institutional training. The successes experienced around the world are not without policy support for the development of clean energy technologies. Enabling policy encourages the development, reliability, cost-effectiveness, and overall understanding of clean energy technologies (Blanco and Rodrigues 2008; Cheng 2005; Couture et al. 2010). Table 7.5 presents some of the policies that enhance the growth of clean energy technologies in developed countries. These policies could be adapted for the development and spread of clean energy technologies in Africa. Bilateral programmes in countries with success stories could also aid the development and diffusion of clean energy technologies in Africa.

Table 7.5 Existing policies for the development of clean energy technologies around the world

Policy	Country	Meaning	Advantages
Investment subsidies (Cheng 2005)	Sweden	Subsidies for investment in clean energy technologies are given on the basis of rated power	Encourages large renewable energy technologies
Fixed feed-in tariffs	Finland, Ireland	Operators are paid a fixed price for every kWh of electricity from clean energy technologies fed into the grid (Couture et al. 2010)	Offsets the high generation cost of clean energy technologies and enhances advance planning. Allows fair price contract with the utility (Erwin 2011)
Fixed premium systems	China, France	Premiums are added to electricity price in a manner to offset the estimated production costs from clean energy technologies relative to conventional alternatives (Bökenkamp et al. 2008)	Offsets high cost of generation from clean energy technologies; it also encourages competitive bidding (Bökenkamp et al. 2008)
Surcharge-funded production subsidy	Germany	Consumers are surcharged on all electricity purchases, and the revenue from the surcharge is distributed to clean energy generators on a per-kilowatt-hour basis for each unit of electricity produced	Incentives for demand-oriented production, increased participation in direct marketing of clean energy technologies (Gawel and Purkus 2013)
Renewable portfolio standards (Ragwitz et al. 2012)	Belgium	Minimum percentage of the electricity produced or sold in the region must come from clean energy technologies	Encourages participation in clean energy industry

(continued)

Table 7.5 (continued)

Policy	Country	Meaning	Advantages
Tradable green certificates	India, Italy	There is an obligation to supply a certain percentage of electricity from clean energy sources (Bökenkamp et al. 2008)	Encourages group of agents in the energy supply chain to have a certain quota of electricity from clean energy technologies in their portfolio
Tendering system	Netherlands, Norway	There is a power purchase agreement for a certain period of years and the price is agreed upon for the period (Gawel and Purkus 2013)	Political risk for the investors is removed in clean energy technologies is removed
Emissions trading schemes	US, India	The scheme sets limit for emission of CO_2. It requires firms to either pay a fine or buy permits from firms that undershoot their targets (Beder 2009)	Clean energy technologies could benefit from a higher emissions allowance price and higher alternative abatement costs
Taxation incentives (Farrell 2008)	US	Lower value-added tax (VAT) rates for electricity from clean technology, favourable depreciation possibilities and exemption from energy or environmental taxes (Gawel and Purkus 2013)	Has good impact on the profitability of clean energy in both short and long run

Criteria for Assessing Clean Technologies

Performance

This is a measure of how well a device/system works. Energy systems are expected to work efficiently, i.e. to convert a given set of inputs to outputs under minimal resource (e.g. energy and fuel) requirements, as such the performance criterion assesses the outputs of a given technology in relation to the inputs or in comparison to other devices. This is typically done in terms of power outputs and energy conversion efficiency (gross or net). The efficiency terms can include electrical efficiency, component efficiencies, fuel efficiency, thermal efficiency, mechanical efficiency, and so on. Some of the factors considered under this criterion can include an understanding of how a technology reduces fuel consumption and ensures emission reduction—a measure of how much a technology is better than the alternatives and can improve the quality of user experience. For example, solar-based (solar-thermal and PV) technologies are promising alternatives to conventional energy sources because of the associated zero cost of fuel; however, the performance of the system depends on the solar energy potential of the location and the efficiency of energy capture, conversion, and storage (Kumar and Sudhakar 2015). The latter is subject to design, installation, operation, and maintenance, and could significantly impair or improve adoption of the technology.

Appropriateness

This is a measure of the suitability and relevance of a given technology to the user's contextual (socio-cultural, economic, technological, environmental, and political) needs. This factor is quite significant for developing economies, as several technologies have failed because of this (Ikejemba et al. 2017). Some of the widely cited reasons include high dependence of technologies on centralised systems (e.g. electric power, supply chain, grid systems, and fuel distribution systems); inadequate availability of fuels; high product complexities; poor product performance and characteristic; poor product design and market models; lack of technical know-how and resource for local manufacture, repairs and maintenance; high product cost; low aesthetic value; poor institutional support and legal framework; and low financial and commercial backings. As such, technologies are required to be designed with the user in mind and special considerations given to socio-cultural, economic, and environmental conditions. Some factors that can be considered include pre-feasibility studies of the needs of the end user rather than preconceived assumptions, ideas, notions, or perceived needs of the end users.

Usability, Accessibility and Availability

This is the extent to which a technology is within the reach of the consumer. This refers to the ways in which the user interacts with the technological innovation: ease of use, how readily available, and whether product is for purchase or hire. It includes the degree of accessibility to maintenance, repair, and part replacement, and extends to the degree to which the user understands and operates the technology. Some of the elements to be considered include the following: Would the technology require a change in user behaviour? Is there sufficient information for the user to access the technology and is this easy to understand? Is specialised training required and how accessible is this to the user? Can the device be operated without some intermediary personnel? Can repair and maintenance be achieved with common hand-held tools or is specialised equipment required? Can parts be replaced with off-the-shelf tools and components and within a relatively short period? Are the users aware that the product exists and is there a sufficient distribution system to support the diffusion of the product? The availability of resources for manufacture can influence the extent to which the technology is produced (individually or in mass quantities) and diffused, provided it is done with relative ease and at minimum cost. For instance, the diffusion of liquefied natural or petroleum gas is limited in Africa due to the poor fuel availability. Modular designs are increasingly used because it eases downtime and facilitates faster service times.

Sustainability

This is a measure of how much a technology contributes to social, economic, and environmental balance, i.e. properties of the system to meet the needs of the user without any adverse impact on natural resources and the needs of future generations (Ashby 2015). Thus, it is important to consider the outputs of the systems including by-products, co-products, and emissions throughout the life cycle of the technology from production to disposal of the unit. The core resources used for producing the technology should not promote natural resource depletion and/or scarcity. The emissions generated from the systems are not expected to cause any adverse effect for the user and the environment. The following can be considered from a life cycle perspective: Can the technology be recycled or reused at the end of its life or does it constitute an environmental burden? Can environmentally non-friendly materials be substituted for environmentally friendly ones? What are the direct burdens associated with the production, use, and disposal of the product, including direct and indirect flows as well as exported emissions? Is the technology economically viable without financial support and can the user afford it? Is the product of any socio-economic benefit? Does the product life cycle cause any additional damage or risk?

Affordability and Willingness to Pay (WTP)

Affordability relates to the cost of the technology to the user, which should be cheap enough for the user to buy or hire. WTP is the extent to which the user is willing to pay to own or hire a technological innovation. It is typically associated with the perceived benefits of the technology. As such, it is important to communicate to the user how the product meets the user's expectations. Some of the elements that can be considered include: additional features that benefit the user (e.g. reduced noise, improved aesthetics, reduced fuel costs, and reduced emissions), availability of consumer financing, subsidies, discounts, incentives, or other financial support that can make the product more affordable, availability of flexible payment systems, and direct and indirect cost savings associated with the use of the product. For instance, there are cost savings associated with the use of solar lanterns in rural areas—hospital costs for respiratory symptoms and diseases; cost for securing kerosene for cooking stoves; and so on.

Others include: (i) resilience, durability, and robustness—the capacity of the technology to work consistently for a reasonable period, even under extreme environmental conditions. This includes extreme ambient conditions such as hot, dry, and dusty, which affect solar cells and thermal collectors; corrosive conditions that affect the durability of materials (usually metals); fuel variability such as high moisture and/or ash content, low calorific value, and the wide varieties of fuel properties that affect engine performance; (ii) safety and reliability—the capacity of a given technology to provide a given set of outputs, without causing risk or harm to the user and user's environment.

Conclusion

This chapter discusses the clean energy sources, potentials, and technologies in Africa, their applications, associated challenges, and mechanisms for promoting their development, transfer, and diffusion. Although the African continent is blessed with an abundance of clean energy resources, they are currently underutilised and thus bear minimal reflection on the continent's essential economic growth and development. The uptake of clean energy technologies in Africa is thus still at the developmental phase in comparison to other parts of the world, particularly in the developed countries, where past investments have been made to harness their clean energy resources. The continent is currently making headway in adopting these technologies to harness its abundant clean energy resources, owing to the increase in the availability of a wide range of low-cost and technology-ready clean energy technologies due to global concern over climate change and other environmental issues. Clean energy technologies are currently being deployed across Africa on various scales, ranging from small household units to monumental regional projects, and for a wide range of applications. Electricity generation stands out as being of critical importance due to its essentiality to the economic growth and development of most African countries, particularly the Sub-Saharan countries, where it is lowest. Despite these promising developments, a time lag is anticipated before these projects are translated into the real economic growth and development that the continent urgently needs. Furthermore, some challenges hindering the development, transfer, and diffusion of these technologies, such as those discussed in the last three sections of this chapter, persist to date. It is thus imperative for African governments and policymakers, in collaboration with other stakeholders, to implement policies that will accelerate the uptake of clean energy technologies and reduce the time lag between clean technology investments and real socio-economic results. Policies, reforms, and initiatives currently being implemented in other parts of the world, such as those presented in this chapter, can serve as a guide for African policymakers. The implementation of effective policies is not only crucial to the utilisation of these abundant resources and uptake of clean energy technologies but also enhances the ability of Africa to fuel its economic growth and self-finance its infrastructural development. Additionally, increasing science and technology funding for African research institutions could increase efficiency and improve clean energy technologies through scientific advances and innovations, which in turn will reduce the cost of production of the technologies. Given the abundance of clean energy resources, incremental growth in clean technologies in Africa, and current and planned policies to mitigate current challenges, the outlook for clean energy technologies in Africa looks promising.

References

AFREPREN/FWD. 2002. AFREPREN Data Base, 2000. Energy Data Reference Handbook VI. African Energy Policy Research Network (AFREPREN/FWD). Nairobi, Kenya.

Africa Progress Panel. 2014. *Grain Fish Money: Financing Africa's Green and Blue Revolutions: Africa Progress Report 2014*. Geneva, Switzerland.

———. 2015. *Power, People, Planet: Seizing Africa's Energy and Climate Opportunities*. Africa Progress Report 2015. Geneva, Switzerland.

Africa-EU Energy Partnership (AEEP). 2016. *Africa-EU Energy Partnership Status Report Update: 2016 A Mid-Term Report on Progress, Achievements and Future Perspectives*. European Union Energy Initiative Partnership Dialogue Facility, Eschborn.

African Clean Energy. 2017. Technology—African Clean Energy. Retrieved from http://www.africancleanenergy.com/technology

Albæk, M. 2015. The Answer Is Blowing in the Wind. *Brunswick Review* 9: 60–62.

Ashby, M.F. 2015. *Materials and Sustainable Development*. Waltham: Butterworth-Heinemann.

Atteridge, A., M. Heneen, and J. Senyagwa. 2013. *Transforming Household Energy Practices Among Charcoal Users in Lusaka, Zambia: A User-Centred Approach*. Working Paper No. 2013–04. Stockholm Environment Institute, Sweden.

Ayodele, T.R., and A.S.O. Ogunjuyigbe. 2015. Mitigation of Wind Power Intermittency: Storage Technology Approach. *Renewable and Sustainable Energy Reviews* 44: 447–456.

Ayodele, T.R., A. Jimoh, J.L. Munda, and A.J. Tehile. 2012. Challenges of Grid Integration of Wind Power on Power System Grid Integrity: A Review. *International Journal of Renewable Energy Research* 4 (2): 618–626.

Balan, V. 2014. Current Challenges in Commercially Producing Biofuels from Lignocellulosic Biomass. *ISRN Biotechnology 2014*, 463074. https://doi.org/10.1155/2014/463074.

Barría, P.R.M. 2016. Indoor Air Pollution by Particulate Matter from Wood Fuel: An Unresolved Problem. *Environment Pollution and Climate Change* 1: 104. https://doi.org/10.4172/2573-458X.1000104.

Beder, S. 2009. Token Environmental Policy Continues in Australia. *Pacific Ecologist* 18: 45–48.

Belward, A., B. Bisselink, K. Bódis, A. Brink, J.-F. Dallemand, A. de Roo, T. Huld, F. Kayitakire, P. Mayaux, M. Moner-Girona, H. Ossenbrink, I. Pinedo, H. Sint, J. Thielen, S. Szabó, U. Tromboni, and L. Willemen. 2011. Renewable Energies in Africa—Current Knowledge. JRC Scientific and Technical Reports JRC 67752. Luxembourg, European Commission.

Blanco, M.I., and G. Rodrigues. 2008. Can the Future EU ETS Support Wind Energy Investments? *Energy Policy* 36: 1509–1520.

Bökenkamp, G., O. Hohmeyer, D. Diakoulaki, C. Tourkolias, R. Porchia, X. Zhu, K.T. Jakobsen, and K. Halsnæs. 2008. WP 9 Report on Policy Assessment of Instruments to Internalise Environment Related External Costs in EU Member States, via Promotion of Renewables. *Cost Assessment for Sustainable Energy Market (CASES)* Project No 518294 SES6: 1–101. Flensburg.

Bozeman, B., J. Hardin, and A.N. Link. 2008. Barriers to the Diffusion of Nanotechnology. *Economics, Innovation and New Technology* 17 (7–8): 749–761.

Cheng, A.Y.C. 2005. *Economic Modeling of Intermittency in Wind Power Generation*. Civil and Environmental Engineering. Massachusetts Institute of Technology, USA, 1–61.

Couture, T.D., K. Cory, C. Kreycik, and E. Williams. 2010. *A Policy Maker's Guide to Feed in Tariff Policy Design*. Technical Report NREL/TP-6A2-44849 (pp. 1–144). National Renewable Energy Laboratory, Colorado.

Danso-Boateng, E., R.G. Holdich, G. Shama, A.D. Wheatley, M. Sohail, and S.J. Martin. 2013. Kinetics of Faecal Biomass Hydrothermal Carbonisation for Hydrochar Production. *Applied Energy* 111: 351–357.

Davis, F.D. 1989. Perceived Usefulness, Perceived Ease of Use and User Acceptance of Information Technology. *MIS Quarterly* 13 (3): 319–340.

De Civita, M.D., and K. Dasgupta. 2007. Using Diffusion of Innovations Theory to Guide Diabetes Management Program Development: An Illustrative Example. *Journal of Public Health* 29 (3): 263–268.

Dearing, J.W. 2009. Applying Diffusion of Innovation Theory to Intervention Development. *Research on Social Work Practice* 19 (5): 503–518.

Deepchand, K. 2002. Promoting Equity in Large-Scale Renewable Energy Development: The Case of Mauritius. *Energy Policy* 30 (11–12): 1129–1142.

Dibra, M. 2015. Rogers Theory on Diffusion of Innovation—the Most Appropriate Theoretical Model in the Study of Factors Influencing the Integration of Sustainability in Tourism Businesses. *Procedia—Social and Behavioral Sciences* 195: 1453–1462.

Dillon, A., and M.G. Morris. 1996. User Acceptance of Information Technology: Theories and Models. *Annual Review of Information Science and Technology* 31: 3–32.

Eberhard, A., K. Gratwick, E. Morella, and P. Antmann. 2016. *Independent Power Projects in Sub-Saharan Africa: Lessons from Five Key Countries*. Directions in Development. Washington, DC: World Bank.

Environmental Protection Agency (EPA). 2011. *Assessing the Multiple Benefits of Clean Energy: A Resource for States*. U.S. Environmental Protection Agency EPA-430-R-11-014. Revised September 2011.

Environmentally Sound Technologies in Africa. n.d. United Nations Economic Commission for Africa, Addis Ababa. Retrieved from https://sustainabledevelopment.un.org/content/documents/1247eca.pdf

Erwin, D. 2011. The Feed in Tariff Model. In *Promoting Renewable Energy Development: An Introductory Workshop for Energy Regulators*, 1–25. Nairobi.

Farrell, J. 2008. *Minnesota Feed-in Tariff Could Lower Cost, Boost Renewables, and Expand Local Ownership.* Minneapolis: Institute for Local Self Reliance, New Rules Project.

Feron, S. 2016. Sustainability of Off-Grid Photovoltaic Systems for Rural Electrification in Developing Countries: A Review. *Sustainability* 8 (12): 1326. https://doi.org/10.3390/su8121326.

Flamos, A. 2010. The Clean Development Mechanism—Catalyst for Wide Spread Deployment of Renewable Energy Technologies? or Misnomer? *Environment Development & Sustainability* 12: 89–102.

Floriš Wind-Powered Water Pumps—Floriš Water Pumps. 2016. Retrieved from http://pfloris.com/floris-wind-powered-water-pumps/

Fullerton, D.G., N. Bruce, and S.B. Gordon. 2008. Indoor Air Pollution from Biomass Fuel Smoke Is a Major Health Concern in the Developing World. *Transactions of the Royal Society of Tropical Medicine and Hygiene* 102 (9): 843–851.

Garcia, Z., J. Nyberg, and S.O. Saadat. 2006. *Agriculture, Trade Negotiations and Gender.* Food and Agriculture Organization of the United Nations, Rome.

Gawel, E., and A. Purkus. 2013. Promoting Market and System Integration of Renewable Energies Through Premium Schemes: A Case Study of the German Market Premium. *Energy Systems in Transition: Inter- and Transdisciplinary Contributions*, 1–19. Karlsruhe, Germany.

Global Alliance for Clean Cookstoves. 2014. *World Energy Resources: Bioenergy 2014.* World Energy Council, London.

———. 2017. *World Energy Resources: Bioenergy 2017.* World Energy Council, London.

GlobalData. 2015. *Power Generation Technologies Capacities, Generation and Markets Database.* London: GlobalData.

Hermann, S., A. Miketa, and N. Fichaux. 2014. *Estimating the Renewable Energy Potential in Africa: A GIS-Based Approach.* IRENA-KTH Working Paper, International Renewable Energy Agency, Abu Dhabi.

Hornor, C., and M. van Gerven. 2015. *Renewable Energy in Sub-Saharan Africa: Opportunities and Challenges.* American Council on Renewable Energy (ACORE), Washington, DC.

Hydropower and Dams. 2014. *Hydropower and Dams World Atlas.* Wallington: Aqua-Media International.

IEA. 2014. *Africa Energy Outlook: A Focus on Energy Prospects in Sub-Saharan Africa*. World Energy Outlook Special Report. OECD/IEA, Paris.

————. 2017. Energy Access Outlook 2017: *From Poverty to Prosperity*. World Energy Outlook Special Report. OECD, Paris.

Ikejemba, E.C.X., P.B. Mpuan, P.C. Schuur, and J. van Hillegersberg. 2017. The Empirical Reality & Sustainable Management Failures of Renewable Energy Projects in Sub-Saharan Africa (Part 1 of 2). *Renewable Energy* 102: 234–240.

Instove. 2017. Cookstoves, InStove. Retrieved from http://www.instove.org/cookstoves

International Energy Agency (IEA). 2006. Energy for Cooking in Developing Countries. In *World Energy Outlook 2006*, 419–445. Paris: OECD/IEA.

International Renewable Energy Agency (IRENA). 2012. *Prospects for the African Power Sector: Scenarios and Strategies for Africa Project*. Abu Dhabi: IRENA.

IRENA. 2013. *Southern African Power Pool: Planning and Prospects for Renewable Energy*. Abu Dhabi: IRENA.

————. 2014. *Global Bioenergy Supply and Demand Projections: A Working Paper for REmap 2030*. Abu Dhabi: IRENA.

————. 2015a. *Africa 2030: Roadmap for a Renewable Energy Future*. Abu Dhabi: IRENA.

————. 2015b. *Renewable Energy Capacity Statistics 2015*. Abu Dhabi: IRENA.

————. 2016. *Solar PV in Africa: Costs and Markets*. Abu Dhabi: IRENA.

————. 2017. *Renewable Capacity Statistics 2017*. Abu Dhabi: IRENA.

Iwayemi, A. 1998. *Energy Sector Development in Africa*. Working Paper 43. African Development Bank, Abidjan.

Jones, L.E. 2017. *Renewable Energy Integration: Practical Management of Variability, Uncertainty, and Flexibility in Power Grids*. Oxford: Academic Press.

Karekezi, S., and J. Kimani. 2002. Status of Power Sector Reform in Africa: Impact on the Poor. *Energy Policy* 30 (11–12): 923–945.

Karekezi, S., and W. Kithyoma. 2003. Renewable Energy Development. *Renewable Energy in Africa: Prospects and Limits. The Workshop for African Energy Experts on Operationalizing the NEPAD Energy Initiative*. AFREPREN, Dakar, Senegal.

Karekezi, S., and T. Ranja. 1997. *Renewable Energy Technologies in Africa*. Oxford: ZED Books and AFREPREN.

Kayembe, J.M., F. Thevenon, A. Laffite, P. Sivalingam, P. Ngelinkoto, C.K. Mulaji, J.-P. Otamonga, J.I. Mubedi, and J. Poté. 2018. High Levels of Faecal Contamination in Drinking Groundwater and Recreational Water Due to Poor Sanitation, in the Sub-Rural Neighbourhoods of Kinshasa, Democratic Republic of the Congo. *International Journal of Hygiene and Environmental Health*. https://doi.org/10.1016/j.ijheh.2018.01.003.

Kirkland, K., and D. Sutch. 2009. *Overcoming the Barriers to Educational Innovation—A Literature Review*. Bristol: Futurelab.

Kumar, B.S., and K. Sudhakar. 2015. Performance Evaluation of 10 MW Grid Connected Solar Photovoltaic Power Plant in India. *Energy Reports* 1: 184–192.

Laffite, A., P.I. Kilunga, J.M. Kayembe, N. Devarajan, C.K. Mulaji, G. Giuliani, V.I. Slaveykova, and J. Poté. 2016. Hospital Effluents Are One of Several Sources of Metal, Antibiotic Resistance Genes, and Bacterial Markers Disseminated in Sub-Saharan Urban Rivers. *Frontiers in Microbiology* 7: 1128. https://doi.org/10.3389/fmicb.2016.01128.

LaMorte, W.W. 2016. Diffusion of Innovation Theory. Boston University School of Public Health. Retrieved from http://sphweb.bumc.bu.edu/otlt/MPH-Modules/SB/BehavioralChangeTheories/BehavioralChangeTheories4.html

Lee, Y.-H., Y.-C. Hsieh, and C.-N. Hsu. 2011. Adding Innovation Diffusion Theory to the Technology Acceptance Model: Supporting Employees' Intentions to Use E-Learning Systems. *Educational Technology & Society* 14 (4): 124–137.

Legros, G., I. Havet, N. Bruce, and S. Bonjour. 2009. *The Energy Access Situation in Developing Countries: A Review Focusing on the Least Developed Countries and Sub-Saharan Africa*. World Health Organization, United Nations Development Programme. New York.

Lighting Africa. 2018. Program Impact as of June 2017. Retrieved from www.lightingafrica.org/about/our-impact/. Accessed 23 Feb 2018.

Liu, L., H.L. Johnson, S. Cousens, J. Perin, S. Scott, J.E. Lawn, I. Rudan, H. Campbell, R. Cibulskis, M. Li, C. Mathers, and R.E. Black. 2000. Global, Regional, and National Causes of Child Mortality: An Updated Systematic Analysis for 2010 with Time Trends Since 2000. *Lancet* 379 (9832): 2151–2161.

Liyanage, C., T. Elhag, and T. Ballal. 2016. Establishing a Connection Between Knowledge Transfer and Innovation Diffusion. *Journal of Knowledge Management Practice* 13 (1). http://www.tlainc.com/articl290.htm.

Masud, J., D. Sharan, and B.N. Lohani. 2007. *Energy for All: Addressing the Energy, Environment and Poverty Nexus in Asia.* Manila: Asian Development Bank.

Mazurkiewicz, A., and B. Poteralska. 2017. Technology Transfer Barriers and Challenges Faced by R&D Organisations. *Procedia Engineering* 182: 457–465.

McCrum, A. 2015. Innovative Clean Energy Technologies—From Concept to Market. Retrieved from https://cleantechnica.com/2015/07/01/innovative-clean-energy-technologies-from-concept-to-market/

National Renewable Energy Lab (NREL). 2017. Concentrating Solar Power Projects in Morocco. Retrieved from https://www.nrel.gov/csp/solarpaces/by_country_detail.cfm/country=MA

Nelson, B.D., P.S. Azzopardi, R. Ahn, P.K. Drain, E. Peacock-Chambers, J. Evert, N. Martineau, V.B. Kerry, and R.E. Pust. 2015. Essential Clinical Global Health: A Multi-National Collaboration Develops a Pioneering New 2015 Textbook for Global Health Trainees and Clinicians Working in Resource-Limited Settings. *Annals of Global Health* 81 (1): 43–44.

Nisbet, R.I., and J.M. Collins. 1978. Barriers and Resistance to Innovation. *Australian Journal of Teacher Education* 3 (1): 1–29.

Nnaemeka, V.E., and E.E. Nebedum. 2016. Policies Enhancing Renewable Energy Development and Implications for Nigeria. *Sustainable Energy* 4 (1): 7–16.

Onabanjo, T., A.J. Kolios, K. Patchigolla, S.T. Wagland, B. Fidalgo, N. Jurado, D.P. Hanak, V. Manovic, A. Parker, E. McAdam, and L. Williams. 2016a. An Experimental Investigation of the Combustion Performance of Human Faeces. *Fuel* 184: 780–791.

Onabanjo, T., K. Patchigolla, S.T. Wagland, B. Fidalgo, A. Kolios, E. McAdam, A. Parker, L. Williams, S. Tyrrel, and E. Cartmell. 2016b. Energy Recovery from Human Faeces via Gasification: A Thermodynamic Equilibrium Modelling Approach. *Energy Conversion and Management* 118: 364–376.

Osok, F.O. 2010. *Barriers to the Development and Deployment of Renewable Energy Technologies in Kenya.* MSc Thesis, University of Oslo.

Painuly, J.P., and J.V. Fenhann. 2002. *Implementations of Renewable Energy Technologies—Opportunities and Barriers: Summary of Country Studies.* UNEP Collaborating Centre on Energy and Environment. Roskilde, Denmark.

Platts McGraw Hill Financial. 2015. World Electric Power Plants Database, June 2015. Retrieved from http://www.platts.com/products/world-electric-power-plants-database

Polzin, F. 2017. Mobilizing Private Finance for Low-Carbon Innovation—A Systematic Review of Barriers and Solutions. *Renewable and Sustainable Energy Reviews* 77: 525–535.

Poole, N., and C.P. Buckley. 2006. Innovation Challenges, Constraints and Opportunities for the Rural Poor. IFAD, Background Paper. Retrieved from https://www.ifad.org/documents/10180/346d2bfa-8853-41d5-bcfd-03ff144effd9

Population Reference Bureau (PRB). 2017. 2017 World Population Data Sheet. Retrieved from http://www.prb.org/Publications/Datasheets/2017/2017-world-population-data-sheet.aspx

Ragwitz, M., J. Winkler, C. Klessmann, M. Gephart, and G. Resch. 2012. *Recent Developments of Feed-in Systems in the EU*. A Research Paper for the International Feed-In Cooperation, A Report Commissioned by the Ministry for the Environment, Nature Conservation and Nuclear Safety, 1–16, Fraunhofer.

Ramachandran, V., A.H. Gelb, and M.K. Shah. (2009). *Africa's Private Sector: What's Wrong with the Business Environment and What to Do About It*. Centre for Global Development. Washington, DC: Brookings Institution Press.

Rehfuess, E. 2006. *Fuel for Life—Household Energy and Health. World Health Organisation (WHO)*. Geneva.

Robinson, S. 2009. *Innovation Diffusion and Technology Transfer*. Kingston: Surveillance Studies Centre.

Roman, R. 2004. Diffusion of Innovations as a Theoretical Framework for Telecentres. *Information Technologies and International Development* 1 (2): 53–66.

Samaradiwakara, G.D.M.N., and C.G. Gunawardena. 2014. Comparison of Existing Technology Acceptance Theories and Models to Suggest a Well Improved Theory/Model. *International Technical Sciences Journal* 1 (1): 1–16.

Shemfe, M. B. 2016. *Performance Assessment of Biofuel Production via Biomass Fast Pyrolysis and Refinery Technologies*. PhD Thesis, Cranfield University.

Shemfe, M., S. Gu, and P. Ranganathan. 2015. Techno-economic Performance Analysis of Biofuel Production and Miniature Electric Power Generation from Biomass Fast Pyrolysis and Bio-oil Upgrading. *Fuel* 143 (1): 361–372.

Shen, W., and M. Power. 2017. Africa and the Export of China's Clean Energy Revolution. *Third World Quarterly* 38: 678–697.

SimGas. 2017. Biogas system—SimGas. Retrieved from http://simgas.org/biogas-system/

SolarAid. 2014. SolarAid Impact Report Autumn 2014. London. Retrieved from https://www.solar-aid.org/assets/Uploads/Publications/SolarAid-Impact-Report-2014.pdf

SolarPaces. 2017. South Africa. Retrieved from http://www.solarpaces.org/csp-technologies/csp-potential-solar-thermal-energy-by-member-nation/south-africa/

Somorin, T.O., and A.J. Kolios. 2017. Prospects of Deployment of Jatropha Biodiesel-Fired Plants in Nigeria's Power Sector. *Energy* 135: 726–739.

Stieninger, M., and D. Nedbal. 2014. Diffusion and Acceptance of Cloud Computing in SMEs: Towards a Valence Model of Relevant Factors. In *47th Hawaii International Conference on System Science*, 3307–3316.

Suberu, M.Y., M.W. Mustafa, N. Bashir, N.A. Muhamad, and A.S. Mokhtar. 2013. Power Sector Renewable Energy Integration for Expanding Access to Electricity in Sub-Saharan Africa. *Renewable and Sustainable Energy Reviews* 25: 630–642.

Sy, A. 2015. Africa: Financing Adaptation and Mitigation in the World's Most Vulnerable Region. In *COP21, Paris: What to Expect*, 58–61. Washington, DC: Brookings Institute.

Tenenbaum, D.J. 2008. Food vs. Fuel: Diversion of Crops Could Cause More Hunger. *Environmental Health Perspectives* 116 (6): A254–A257.

The Humane Society of the United States (HSUS). 2009. An HSUS Report: The Implications of Farm Animal-Based Bioenergy Production. HSUS Reports: Farm Industry Impacts on Environment and Human Health. Washington, DC, USA.

Tiyou, T. 2016. The Five Biggest Wind Energy Markets in Africa. *Renewable Energy Focus* 17 (6): 218–220.

UNECA. 2017. Mechanisms to Promote Development and Dissemination of Clean and Environmentally Sound Technologies in Africa. United Nations Economic Commission for Africa, Addis Ababa. Retrieved from https://sustainabledevelopment.un.org/content/documents/1247eca.pdf

UNIDO. 2017. Facilitation Mechanisms to Promote the Development, Transfer and Dissemination of Clean and Environmentally Sound Technologies. United Nations Industrial Development Organization. Retrieved from https://sustainabledevelopment.un.org/content/documents/1293unido.pdf

Union of Concerned Scientists (UCSUSA). 2017. Barriers to Renewable Energy Technologies. Retrieved from http://www.ucsusa.org/clean_energy/smart-energy-solutions/increase-renewables/barriers-to-renewable-energy.html#.WgwYGVu0Pcs

United Nations Department of Economic and Social Affairs (UN DESA). 2007. *Small-Scale Production and Use of Liquid Biofuels in Sub-Saharan Africa: Perspectives for Sustainable Development*. Background Paper No. 2. DESA/DSD/2007/2. UN DESA, New York.

United Nations Environmental Programme and European Patent Office (UNEP and EPO). 2013. Patents and Clean Energy Technologies in Africa. UNEP and EPO. Retrieved from http://www.epo.org/clean-energy-africa

United States Energy Association. 2017. East Africa Geothermal Partnership (EAGP). Retrieved from http://www.geo-energy.org/EastAfrica/EAGP.aspx

Utria, B.E. 2004. Ethanol and Gelfuel: Clean Renewable Cooking Fuels for Poverty Alleviation in Africa. *Energy for Sustainable Development* 8 (3): 107–114.

Waruru, M. 2014. East African Countries Move to Adopt Renewable Energy Technologies. Renewable Energy World. Retrieved from http://www.renewableenergyworld.com/articles/2014/12/east-african-countries-move-to-adopt-renewable-energy-technologies.html

WHO and the United Nations Children's Fund (UNICEF). 2017. Progress on Drinking Water, Sanitation and Hygiene: 2017 Update and SDG Baselines. WHO/UNICEF Joint Monitoring Programme for Water Supply, Sanitation and Hygiene. Geneva, Switzerland.

Wojuola, R.N., and B.P. Alant. 2017. Public Perceptions About Renewable Energy Technologies in Nigeria. *African Journal of Science Technology Innovation and Development* 9: 399–409.

World Bank. 2017a. Clean Energy Transition Will Increase Demand for Minerals, Says New World Bank Report. Press Release NO: 005. Retrieved from http://www.worldbank.org/en/news/press-release/2017/07/18/clean-energy-transition-will-increase-demand-for-minerals-says-new-world-bank-report

———. 2017b. Making Renewable Energy More Accessible in Sub-Saharan Africa. Retrieved from http://www.worldbank.org/en/news/feature/2017/02/13/making-renewable-energy-more-accessible-in-sub-saharan-africa

World Health Organization (WHO). 2010. Barriers to Innovation in the Field of Medical Devices. Background Paper 6. August 2010. WHO/HSS/EHT/DIM/10.6. Geneva: World Health Organization. Retrieved from http://www.who.int/iris/handle/10665/70457

Wu, G.C., R. Deshmukh, K. Ndhlukula, T. Radojicic, J. Reilly-Moman, A. Phadke, D.M. Kammen, and D.S. Callaway. 2017. Strategic Siting and

Regional Grid Interconnections Key to Low-Carbon Futures in African Countries. *Proceedings of the National Academy of Sciences* 114 (15): E3004–E3012.

Zhang, X., P. Yu, J. Yan, and I.T.A.M. Spil. 2015. Using Diffusion of Innovation Theory to Understand the Factors Impacting Patient Acceptance and Use of Consumer E-health Innovations: A Case Study in a Primary Care Clinic. *BMC Health Services Research* 15 (1): 71. https://doi.org/10.1186/s12913-015-0726-2.

8

Renewable Energy in Africa: Policies, Sustainability, and Affordability

Augustine O. Ifelebuegu and Peter Ojo

Introduction

Energy is an essential requirement and central building block for economic development, growth, and for maintaining a modern lifestyle. It plays a critical role in the production of goods and services such as healthcare, water supply, and education, and empowers the communities to increase their productivity (Kolawole et al. 2017). Providing access to affordable energy services remains a key objective of every nation. Over 1.3 billion people globally remain without access to electricity. Almost all these populations live in developing countries, particularly countries in Sub-Saharan Africa (SSA). The African Development Bank (AfDB) estimates that over 645 million Africans remain without access to electricity (AfDB 2010). In the last two decades, there has been significant attention and effort geared towards energy development to support rural development in Africa. Although the rate of electrification has barely kept up with the growth rate

A. O. Ifelebuegu (✉) • P. Ojo
Coventry University, Coventry, UK
e-mail: aa5876@coventry.ac.uk

© The Author(s) 2019
S. Adesola, F. Brennan (eds.), *Energy in Africa*,
https://doi.org/10.1007/978-3-319-91301-8_8

of the population in the region, since 2000 energy usage in SSA has grown by up to 45% (IEA 2015). The energy consumption in SSA is currently 181 kWh per annum compared with 6500 kWh and 13,000 kWh in Europe and the USA respectively. This lack of adequate access to electricity is costing the continent 2–4% of its annual gross domestic product. Although the total power generation capacity in Africa is expected to quadruple to about 385 gigawatts (GW) by 2040 according to the International Energy Agency (IEA), over half a billion people are still expected to remain without electricity. Powering the continent is a necessity in achieving poverty reduction and sustained economic development in Africa.

A vast majority of the energy used globally has been sourced from the burning of fossil fuels. Fossil fuel sources are in decline and not sustainable. Also, the development of fossil fuel power plants is often very expensive and time-consuming when developing large-scale grid connections. Renewable energy is now more attractive, particularly where mini-grids and decentralised systems are needed to meet the energy requirements of rural communities. Renewable energy generation is growing by the day. In 1987 the World Commission on Environment and Development described renewable energy as the global energy structure of the twenty-first century, and since the early 1990s, there has been an increase in the development and spread of renewable energy sources such as solar, wind, biomass, and hydropower. The contribution of renewable energy to the global energy mix, although growing in significance, still remains at about 10% of net energy production worldwide.

Renewable energy is energy obtained from infinite and self-replenishing sources that occur in the natural environment. It is produced from sources that are regenerative and hence offer the world a sustainable alternative to fossil fuel-based energy, which is in constant decline. Renewable energy development and adoption will, therefore, put our energy generation and use, and hence the environment, on a more sustainable footing in an increasingly carbon-constrained world.

The opportunity to exploit renewable energy from hydro, wind, biomass, waves, and solar has fascinated people over the years. The globally installed capacity for renewable energy is now over 2 million megawatts (MW). Of these, Africa has an installed capacity of about 38,000 MW (IRENA 2015b), despite having a huge renewable energy source compared with the rest of the world. Countries in the North African region are richly endowed with renewables, especially solar and to some extent wind, and have been reported among the world's best direct normal radiation levels (Griffiths 2017). Renewable energy will make a major contribution to meeting Africa's energy needs in a sustainable way. As part of the drivers in alleviating the energy crisis in the African continent, the Sustainable Energy for All Initiative, among others, is a global initiative reported by the United Nations Secretary-General in 2012 with a goal of providing worldwide access to modern energy services by 2030. The initiative's three interlinked objectives to be achieved by 2030 are to ensure universal access to modern energy services, double the global rate of improvement in energy efficiency, and double the share of renewable energy in the global energy mix. The third objective is a key driver for the deployment of renewable energy into Africa's energy mix. This is particularly important because Africa cannot power its homes and business unless it unlocks its huge renewable energy potentials. Renewable energy is therefore expected to play a significant role in addressing the widespread poor access to electricity in SSA, particularly in the rural areas.

This chapter examines the renewable energy potential of the continent, policy issues, affordability, sustainability, and management challenges. As Africa leads the world in terms of renewable energy potentials, section "Africa's Renewable Energy Resources" examines the various sources of renewable energy and their potentials within the continent. In section "Policies and Regulatory Framework", we discuss the policy and regulatory framework in place to support the actualisation of the targets for renewable energy deployment. Section "Sustainability and Development of Renewables in Africa" examines the issue of sustainability and section "Affordability and Renewable Energy Financing and Management" focuses on the affordability of renewable energy and the various financing opportunities for the development of the renewable energy potential in Africa.

Africa's Renewable Energy Resources

The African continent's huge renewable energy potential fluctuates across its different regions. Whilst solar energy capacity is on the high side across the regions of Africa, hydropower has some coverage towards the southern areas and wind energy has more potential, not just in the southern region or eastern areas but also around the northern regions of Africa. However, due to climate change issues, human activities, and economic development within the African continent, there is a continuing search for a cleaner source of energy as the demand is expected to double by 2030. This makes the study on sustainable deployment of renewables (solar photovoltaics [PV], hydropower, biomass, wind, and geothermal) into the energy mix in the African continent a great necessity.

Solar Resources

Africa is leading the rest of the world in terms of available solar power potential. The continent receives many more hours of sunshine per annum than any other continent on Earth. The eastern Sahara/north-eastern Africa has the world sunshine record, with up to 4300 hours per annum of sunshine (Dunlop 2008). The theoretically, potentially recoverable solar energy in Africa is currently estimated to be in the region of 10 terawatts (TW). Solar energy is recovered mainly using solar PV or concentrated solar power (CSP). The PV systems use solar cells to convert solar energy directly to electricity, while the CSP system uses sunlight to heat fluid which can then be used to generate electricity. CSPs are also referred to as thermal power plants. Both PV and CSP are unique solar energy options but CSP has greater relevance for rural areas.

The reported installed capacity for solar in 2014 was 1334 MW. However, Kenya had a share of 60 MW in 2014 and South Africa reported another 780 MW between 2013 and 2014. The additional proposed sky power project for solar PV installation in Kenya, Nigeria, and Egypt promises to deliver 7 GW by 2020 (IRENA 2015a). Solar energy resources have not been fully explored in the African continent (Adeoti et al. 2001). Currently, Morocco is leading the way in Africa in the deployment of solar PV and CSP technologies. The Moroccan Energy Strategy (MES) for 2020–2030 drives this initiative. The installed renewable energy capacity is projected to rise by 42% by 2020 (Linklaters 2016). In addition, South Africa's Renewable Energy Independent Power Producer Procurement (REIPPP) programme has equally powered a profitable platform for renewable energy funding and about 7 GW of renewable energy is expected to be operational by 2020.

Wind Resources

The estimated wind potential in the African continent is 110 GW. The total connected capacity of wind energy between 2013 and 2014 was 2462 MW. Although Morocco has been identified as the leading country in wind energy generation with a proposed wind power capacity of 2 GW by 2030, South Africa has also shown some remarkable growth with 8.4 GW projected by 2030. In Egypt, a projection of another 7 GW by 2020 has been reported (IRENA 2015d). The Global Data (2015) reports an operational capacity of 21 GW from 140 wind farms by 2020 and another 300 MW in East Africa. An increase in wind energy capacity between 75 GW and 86 GW is projected by 2030. Onshore wind energy resources are expected to be the lowest and most cost-effective option for electricity generation in the future (GWEC 2014).

Hydropower Resources

The hydro resources in Africa have been reported to be underutilised with 350 GW potential untapped (IRENA and IEA-ETSAP 2015). The current total installed capacity in Africa is 28 GW. The capacity of hydropower in central Africa was estimated at 40% of the African continent's hydro resources but East Africa accounted for another 28% (Hydropower and Dams 2014). There are large hydro capacities in the Congo River, Zambezi, Niger, and Nile (IRENA 2015a). The grand Inga project on the Congo River with 40 GW hydropower power generation will make the river one of the major hydropower plants globally (World Bank 2014). The hydropower costs in Africa are as low as $0.03/kWh, while the mean value is $0.10/kWh, but looking at a large-scale hydropower plant, the mean installed cost is $1400/kW (IRENA 2015c).

Geothermal Resources

According to the Geothermal Energy Association (2015), geothermal resources have great potential in East Africa and South Africa with 15 GW capacity. Towards the end of 2014, 606 MW was reported in Africa, but 579 MW was also installed in Kenya. Other capacities include 640 MW from Tanzania and Ethiopia by 2018 (IRENA 2015a). The cost of current projects in East Africa reported in IRENA (2015c) ranged from $2700/kW to $7600/kW and the weighted mean value was $4700/kW. Research data have also reported between 4000 and 7000 MW of geothermal electricity potential unexplored in Kenya (Simiyu and Keller 2000).

Biomass Resources

Almost one-third of SSA is covered by forest, with total forest biomass stock estimated to be 130 billion tonnes as at 2010. The biomass resources can be either biofuels, wood fuels, or biomass waste resources. Sugar and starch crops have found relevance in both biodiesel and bioethanol. East Africa is known for ethanol generation, while South Africa has potentials in plant oil crops. According to IRENA (2014), the challenges of estimating biomass potentials in the African continent exist because of the issues around agricultural food production and land use decrees. The energy content of biomass resources potentially available for conversion into liquid biofuel is estimated to be 4.8 Exajoule (EJ)/year by 2030. However, 3.6 EJ of this energy capacity is linked to crops for ethanol production, with southern Africa accounting for 65% of ethanol potential and another 20% reported in East Africa and most of the others in Central Africa. An insignificant capacity was identified in West and North Africa (IRENA 2014). The total cost of fitting biodiesel plants is typically lower than that for ethanol production but North Africa and the Middle East showed cost values of $0.25/litre/year of production capacity (IRENA 2013).

Benoit (2006) and Karekezi and Kithyoma (2003) identified biomass as the principal renewable option in the African continent but the reported consumable biomass resources comprise domestic waste, forest tree resources, animal residues, crop waste, waste paper materials, wastewater, and industrial biodegradable waste; the exact concentration of these resources in the African continent has not yet been reported. However, recent research studies have shown 30% of agricultural residues accounting for 5000 MW, while 10% of forest wood accounted for 10,000 MW. Ghana's renewable potentials from crop leftovers accounted for 75.20 terajoules (TJ); 697.15 TJ were reported for Nigeria but an untapped 2600 MW lies fallow in Uganda (Daka and Ballet 2011; Dasappa 2011; Felix and Gheewala 2011; Kassenga 1997). Oil palm fruit production with biodiesel renewable exploitation has been reported to be abundant in Nigeria, and Ghana has been reported to house another 320,000 hectares of oil palm plantations (Duku et al. 2011; World Trade Organisation 2007).

The benefits of this form of renewables range from utilisation as a blending agent in petrol—for example, ethanol at levels between 5% and 15% in southern African countries such as Kenya, Malawi, and Zimbabwe (Lerner 2010) and the application in transportation, as in the case of Benin, Burkina Faso, Cote d'Ivoire, Gambia, Ghana, Guinea Bissau, Liberia, Mali, Niger, Nigeria, Senegal, Sierra Leone, Togo, and Uganda, where national liquid biofuel targets for transportation have been established (IRENA 2015d).

Considering the rural context, cooking in Africa is still being carried out using the conventional biomass approach, which generates considerable amounts of fumes and particulate substances with a prolonged adverse health impact on the users. Although the cost of efficient cookstoves is in the range of $5–$10 (Global Alliance for Clean Cookstoves 2014), there still exist challenges to procuring this cooking device for low-income earners. However, subsidies and microfinance schemes with a repayment plan could be an option to deal with this challenge. Another option is the use of charcoal compared with firewood in the rural context but again, over the past 40 years, yearly generation of charcoal has risen at a mean annual rate of 6.3% (FAO 2015). A strategy to push for a sustainable charcoal feedstock sourcing and monitoring system has been made but has not been positively received. The UN-HABITAT (1993) has reported reasonable efficiencies of 10% and 20% using traditional earth kilns for charcoal generation, and 25% and 40% efficiencies with improved metal, brick, and retort kilns.

Policies and Regulatory Framework

The growth of renewable energy in Africa is driven partly by the national and domestic policy frameworks and governance structures in place. Currently, the continent trails the world in government policies that will promote renewable energy adoption. In the World Bank Inaugural Regulatory Indicators for Sustainable Energy report, more than 40% of countries in Africa have barely begun to take policy measures to accelerate access to energy. Ethiopia, Nigeria, and Sudan are of concern, with South Africa, Tunisia, and Morocco identified as the bright spots for a policy framework for the development of energy access (World Bank 2017). Table 8.1 represents some of the current policy initiatives for renewable energy targets, feed-in tariffs (FITs) premium payments, transport obligation and mandate, and heat obligation and mandate for major countries of the African continent.

Besides the individual policy initiatives of the different countries, there are several regional and international initiatives to help the continent achieve access to affordable energy. Africa must consider regional cooperation in energy generation to achieve the continent's goals of electrifying the entire continent. Regional energy generation will provide an optimal economic solution to energy generation and use, because energy is generated where it is most economical. There is, therefore, a need for regional policy direction and initiatives for the development and deployment of renewable energy.

The New Partnership for Africa's Development (NEPAD) early in the new millennium proposed strategic development objectives for meeting the energy needs of the continent in line with the United Nations' Sustainable Development Goals (SDGs). One of the key targets of the SDGs is to improve the availability of energy from 10% to 35% in two decades. Several of the region's economic communities have developed strategies in line with NEPAD. Several regional and international agencies have also been active in developing frameworks for addressing the energy needs of Africa. The World Bank put forward an action plan to make energy available in Africa through the investment framework for clean energy and development (Brew Hammond 2010). Also, recently the AfDB's new deal on energy for Africa has set a target for universal access to

Table 8.1 Renewable Energy Support Policies (REN 21-Renewable energy policy network 2016)

Countries	Renewable energy targets	Feed in Tariffs premium payments	Transport obligation/ Mandate	Heat obligation/ Mandate
Angola		E	E	
Botswana	E			
Libya	E			
Namibia	E			E
South Africa	R	E	E	E
Cabo Verde	R			
Cameroon	E			
Cote d'Ivoire	E			
Egypt	E	R		
Ghana	R	E	E	E
Kenya	R	R		E
Lesotho	E			
Morocco	R	E		
Nigeria	R	E	E	
Sudan	R		E	
Tunisia	R			
Zambia	Q			
Burkina Faso	R			
Ethiopia	R		E	
Gambia	R			
Guinea	E			
Liberia	E			
Madagascar	R			
Malawi	R			E
Mali	E		E	
Mozambique	E		E	
Niger	R			
Rwanda	R	E		
Senegal	R	E		
Tanzania	R	E		
Togo	E			
Uganda	E	E		
Zimbabwe			R	

E Existing National, *Q* New and *R* Revised (one or more policies of this category) electricity across the African continent by 2025 (AfDB 2010). To achieve this goal, they have projected the addition of new 169 GW on-grid generation, 130 million new on-grid connections, and new off-grid generation to add over 75 million new connections. More policy initiatives, particularly at the national level, are required to help Africa, and in particular SSA, to achieve its renewable energy goals and potentials.

As Morocco has been identified as a leading strength in driving renewables in the Middle East and North Africa (MENA), it becomes paramount to cite Morocco's renewable energy target of 42% of total installed capacity by 2020 and 52% by 2030 as a model for sustainable and affordable energy resources in the African continent (Hochberg 2016; Energy and Mines 2016). In realising this renewable obligation by 2020 and 2030, the Moroccan government is driving this through different strategies that include, the renewable energy law 13-09 flagged in 2009 with a key responsibility to develop the legal framework that unlocks the electricity market to a contest of marketability and generation of electrical energy from renewables (Ministry of Energy, Mines and Sustainable Development 2015a, b). This has empowered private units to deploy renewable energy production projects and commercialise energy to huge consumers with available openings to Morocco's international grid linked to Spain and Algeria for energy transportation. This initiative has allowed renewable energy investors to develop their own transmission routes and export the electrical energy generated and it has boosted Morocco's energy mix target by 2760 MW of wind and solar power. The global strategy of competitive tenders (competitive reverse auctions) has also been utilised in driving the deployment of renewable electrical energy in Morocco. This initiative allows independent power producers (IPPs) to bid for the generation of each megawatt hour (MWh) and the lowest bidder is chosen during the auction process. The IPPs and Morocco's state-owned electricity company and renewable market operators have helped to sign agreements to consolidate the wind power purchase agreements (PPAs) enacted during the competitive tender process. The PPAs further empower the renewable energy investors financially and successful bid prices as low as $25/MWh for solar PV and $28/MWH for wind have been achieved.

In South Africa, renewable energy is driven by the government's ZAR1.2 billion/year electrification initiatives to provide all households in

South Africa with electrical energy. The National Electricity Regulator (NER) as a key driver operates from two main platforms, namely, non-grid electrification projects and hybrid mini-grid systems. The "fee-for-service" (FFS) strategy is utilised in South Africa and non-grid service providers are providing electricity to customers in a specific area (United Nations Commission on Sustainable Development 2015). However, energy is provided through solar home systems (SHSs) but the household does not own the equipment; it just pays a service fee per month. Another initiative utilised in South Africa is the Tradable Renewable Energy Certificates (TRECs) in which renewable energy producers receive a TREC, which they can trade nationally or internationally with investors that sign up into the green attitudes (green premium) and this forms a sustainable renewable energy financing initiative that will cut down on government financing capacity (Department of Minerals and Energy 2003). The legal strategy promotes a legalised grid-connection code that monitors the acceptable requirements of connecting to the grid platform.

In the case of Kenya, the country utilises the Energy Act 12 of 2006 renewable energy policy, which was ratified by the Ministry of Energy and Petroleum and is a self-regulating power for renewables. The other regulatory bodies include the Energy Regulatory Commission (ERC), responsible for the endorsement of PPAs and the preparation of national energy plans, such as enlargement of indigenous producers of renewable energy, and for providing further motivation for conventional renewable energy sources such as bio digesters, solar, and hydro turbines (IEA 2016). The Energy Act also drives initiatives such as bioenergy cogeneration (combined heat and power systems) for boosting sugar mill operations in Kenya.

Sustainability and Development of Renewables in Africa

The concept of sustainable development evaluates the ways of meeting present human needs without hindering the future generations' rights and access to the same resources (Ifelebuegu 2013; UN 2018). In meeting the needs of humans, a great deal of naturally occurring resources are spent and most often unsustainably, for the provision of infrastructures such as potable water supply, energy supply, and transportation. Renewable energies are produced from the continuous outflow of energy in the natural environment often with zero emissions and are therefore high on the global sustainability agenda. Consequently, there has been a gradual shift of renewable energy from the fringe to the mainstream of sustainable development in recent years, particularly in the Global South. This is mainly because of the contribution of renewable energy expansion to climate change mitigation, rural and community development, and energy security (Martinot et al. 2002). According to Kammen (2015), the policy drive towards renewable energy development has been advocated on the grounds of environmental sustainability. It is a major consideration in evaluating the long-term viability of any energy source.

Following the adoption of the SDGs, which focus on sustainability targets such as alleviating scarcity and starvation, cutting down on carbon emission, and promoting good wellbeing of the people, the acceleration of renewable energy will play a vital role in meeting these targets (United Nations 2015). The need for water, energy, transportation, and all that impacts on life's comforts constantly puts great pressure on these natural resources. The growing African population does not correspond to the economic growth and index, and therefore shows that there exists an unsustainable competition for these naturally occurring resources and a clarion call for sustainable solutions, with the emphasis more on energy needs. The drive of renewable energy has a strong collaborative impact on the SDGs for several reasons: it delivers *energy*, which provides the *platform* for delivering many *goals*. Having a closer look at the African continent, sustainability of energy is a determining factor not just for cutting down the scarcity level in African society, mechanising agricultural development, and empowerment in knowledge capacity but also for increasing the amount of available goods and services per population equivalent in the continent (Daka and Ballet 2011; Eggoh et al. 2011; Barnes et al. 2014; Mushtaq et al. 2009). Driving renewable energy obligation will allow African countries to make savings on carbon emissions from conventional fossil fuel resources.

Deichmann et al. (2011) reported that although the SDGs promote a cleaner energy option at a reasonable and cost-effective price in both urban and rural settlements in African countries, the renewable energy strategy taking the lead is solar resources, due to its cost effectiveness in the renewable energy market. Hydro resources have also been reported in the IPCC report as being inexpensive and that a potential 92% of available resources has not yet been harnessed (IPCC 2011). Renewable energy is a cleaner energy option that can significantly reduce reliance on fossil fuel. This cleaner option will mitigate fluctuations in the fuel cost and empower job creation, and overall through the "Clean Development Mechanism" and Renewable Energy Obligation Certificates, drive the sustainability of renewable energy deployment in the energy mix and boost the economy (Pegels 2010).

Affordability and Renewable Energy Financing and Management

Affordability

The end-user prices of energy vary significantly across SSA. The difference is dependent on the relative ease of supply and the extent of government control. Most often, governments subsidise energy costs, which in some way have hindered private investment, making affordability a continuing challenge. Affordability of renewable energy has remained a major issue, particularly in the Global South. The term is used to describe the possibility to purchase a subsistence level of consumption (typically 50 kWh in SSA) without spending more than a given share of the household budget (Winkler et al. 2011). Renewable energy in its raw forms is generally free and abundant in nature. However, there is a significant cost required to collect, process, and transport the energy to the point of use. The levelised cost of energy (LCOE), which is used to describe the net present value of the unit cost of electricity over the lifetime of a generating asset, is currently higher than the cost of conventional energy generation, as illustrated in Table 8.2.

However, in the last decade, the LCOE has been moving towards cost competitiveness with fossil fuel sources. Between 2009 and 2014 there was a 75% drop in the cost of solar PV modules (IRENA 2015a). A decade ago, the cost of generating energy from solar was over $0.6 per kWh. Today, the LCOE for solar and wind are in the region of $0.04 and $0.1

Table 8.2 US-DOE
Annual Energy Outlook
2017

Power plant type	Cost $/kW-hr
Coal	$0.11–$0.12
Natural gas	$0.053–$0.11
Nuclear	$0.096
Wind	$0.044–$0.20
Solar PV	$0.058
Solar thermal	$0.184
Geothermal	$0.05
Biomass	$0.098
Hydro	$0.064

Table 8.3 Levelised Cost of Energy (LCOE) for renewables and fossil fuels (2014 US$/kWh) (Linklaters 2016)

Renewable energy and fossil fuels	Global LCOE	Average LCOE for Africa
Biomass	0.04–0.25	0.06
Geothermal	0.04–0.11	–
Hydro	0.03–0.024	0.11
Solar PV	0.06–0.425	0.18
Concentrating solar power (CSP)	0.16–0.31	
Wind offshore	0.09–0.30	
Wind onshore	0.025–0.17	0.09
Fossil fuel	0.05–0.13	

per kWh respectively compared with the cost of electricity generation from fossil fuels, which is in the range of $0.05–$0.13 per kWh (Table 8.3). According to a report by the World Economic Forum, renewable energy has reached a tipping point with solar and wind energy generation becoming very competitive in comparison with traditional generation methods. According to the report, the LCOE for solar and wind is now on a par with new fossil fuel capacity in more than 30 countries of the world. According to IRENA (2015b), further falls in the LCOE are expected for these and other renewable sources. It is projected that all renewable energy sources will be cost competitive with fossil fuels by 2020. Despite the reductions in LCOE renewable energy, solar PV in particular remains an expensive option for Africa due mainly to the high cost of project financing, which erodes value because of the relatively high upfront capital cost of installation (Ondraczek et al. 2015).

There is currently a financial necessity for the energy structure in the African continent. A study by the World Bank estimates a renewable energy financing requirement of $43 billion per year for energy in Africa (Briceno-Garmendia et al. 2008). A policy evaluation for the AfDB by Duarte et al. (2010) also reported a comparable amount of $41 billion (UNEP 2012). In another World Bank review, Eberhard et al. (2011) estimated $40.8 billion per year needs to be spent in the power sector in Africa, but currently $11.6 billion is spent and an additional $8.24 billion has been projected. The new projected figure can be actualised by tackling key issues such as utility inadequacies, the under-pricing of power, and poor budget implementation. Funds gathered from taxes and utility charges account for 80% of total expenditure on energy infrastructure in Africa (Africa Progress Panel 2015). This includes an annual amount of $4 billion since 2010 from secluded participation in infrastructure (Gutman et al. 2015). The balance has been financed by international donors. There are basically two major drivers for financing renewable energy in Africa: development finance institutions (DFIs) and climate funds.

Development Finance Institutions (DFIs)

In tackling the climate change challenge, the World Bank has put in place financial resources to drive and manage this project. The World Bank has pursued this project as the green bonds initiative with a clear target of financing the climate change mitigation project. The green bonds initiative's financial base includes Climate Investment Funds (CIFs) and the Strategic Climate Fund (SCF) alongside the Carbon Partnership Facility (CPF). The World Bank provides support of over $2 billion per year for renewable energy projects in Africa with a financial support platform

Table 8.4 Funding approved (US $million) for the energy sector in Africa by the World Bank (Gujba et al. 2012)

Funding source	IBRD and IDA others 2009–2014		IBRD and IDA others 2014	
Renewables and electricity	11,567.10	65.16	2080.84	13.50
Oil and gas	1936.66	5.38	1936.66	0.85

ranging from the International Bank for Reconstruction and Development (IBRD) to the International Development Association (IDA) (Table 8.4).

Known to provide financial portfolios to African nations seeking to deploy renewables in the energy mix is the AfDB (AfDB 2010). The easily explored financial portfolio has a capital base of $2 billion for financing renewable energy deployment in African countries. The AfDB further assists the CIF with a financial capacity of $625 million on a yearly basis and other platforms of support include enabling access to the Global Environment Fund (GEF), the Green Facility for Africa (GFA), and two risk-guarantee products, namely, AfDB Partial Risk Guarantees (PRGs) and the African Development Fund (ADF) PRGs, which encourage the private sector to participate in renewable projects in Africa.

Climate Funds

Renewable energy development in Africa was initially driven mostly by the need to cut down on greenhouse gas (GHG) effects directly linked to climate change (UNFCCC 1997). The mitigation of GHG effects is driven on the platform of the clean development mechanism (CDM) strategy involving the deployment of GHG emission reduction techniques in exchange for GHG emission reduction credits associated with CDM projects (Abanda et al. 2012; UNFCCC 1997). On the other hand, international finances are put together to finance projects designed to mitigate climate change and drive the deployment of renewable energy. Donors are adopting a results-based financing mechanism which reduces the risk for the donors.

There are numerous climate change-related international funds designated for the development of renewable energy projects, particularly in the Global South, and SSA in particular. The Green Climate Fund (GCF) is planned to become one of the key funding vehicles for renewable energy development. It was established by the Cancun Agreement in 2010 to finance climate change mitigation activities. Before the establishment of this financial vehicle, a total of 15 climate funds were already available in Africa (Afful-Koomson 2015) and these financial resources are sponsored by DFIs through multifaceted contributors. The CIF is an example of a DFI that is sponsored by the World Bank and the regional development banks such as AfDB. The CIF is composed of two climate funds: the Clean Technology Fund (CTF) and the Strategic Climate Fund (SCF). In Africa about 42% of CTF sponsorship has been utilised and among the examples of multidimensional benefactors are the Global Environment Facility Trust Fund (GEFTF) and the Global Energy Efficiency and Renewable Energy Fund (GEEREF). The GEFTF powers emerging and developing countries in realising the United Nations Framework Convention on Climate Change (UNFCCC) target into alleviating climate change. The GEEREF is supported by the European Union, Germany, and Norway and operates as a Public Private Partnership (PPP) with the target of providing financial aid to small and medium-sized enterprises (SMEs) that participate in energy efficiency and renewable energy, especially in a developing context (UNEP 2012; Gujba et al. 2012). About $26.96 million has been used to finance two private equity funds in SSA for delivering clean energy in developing countries.

Financing Capacity

The total amount of yearly financial resource utilised for funding renewable energy in Africa is difficult to evaluate. Table 8.4 represents some of the amounts involved. In 2014 over $2 billion of funding was used for renewable energy development in Africa. Afful-Koomson (2015) reported that all funds together agreed on 492 climate change mitigation projects in Africa with a value of $3.5 billion between 2003 and 2013. Of the approved amount, 13% was disseminated for renewable energy projects, while 46% was used for mitigation. The average cost of a project is $3.71 million and most approved projects obtained less than $10 million.

The magnitude of funding required for improving energy generation capacity varies with size. According to the IEA (IEA 2012), a typical solar PV system costs between $1.4/W and $3.3/W and the majority of power plants in the continent are larger than 50 MW; countries such as Ghana, Nigeria, and South Africa operate power plants with a capacity in excess of 1000 MW. However, about 80 solar PV power plants with a size of at least 50 MW have been built globally (though none of them in Africa) but the Nzema Solar Park in Ghana is planned to have a capacity of 155 MW by 2015 (Eshun and Amoako-Tuffour 2016), indicating the potential commercial viability of large-scale PV plants in Africa. The climate funds (CF) have been known to be a key source of most financial support for renewable projects in Africa. A new borrowing strategy has been reported to be more promising. Discounted loans with low interest rates tend to boost investors in renewable energy projects in Africa (Afful-Koomson 2015). FITs are also a vital and efficient policy strategy to empower the sustainable deployment of renewable energy technology globally because this strategy is known to drive sustainable prices for electrification from renewable energy development projects for a fixed time window (Couture and Gagnon 2010). On the other hand, the strategy suffers from challenges in the African continent because of the regulatory, financial, and technical limitations which have made the FITs market diminish (Heinrich Boell Foundation 2013; Rickerson et al. 2013). However, this challenge was averted in Tanzania as an initiative for sustainable renewable energy deployment was implemented in the case study of standardised power purchase agreements (SPPAs), which was linked to their FIT policy in order to suit the socio-economic context of Tanzania. The SPPAs were established to assist the procurement of grid-connected renewable energy power and off-grid renewable energy power and the associated electric energy between the purchasers and small power project manufacturers (Heinrich Boell Foundation 2013).

Conclusion and Future Outlook

Africa has a huge untapped renewable energy potential which varies in types and sizes in the geographical regions. The expected energy demand of Africa is expected to double by 2030 due to rapid economic growth, changing lifestyles, and the need for reliable modern energy access. The answer to these quests for reliable modern energy access and broadening of the electricity supply lies in the deployment of renewable energy. The clarity of policy and renewable energy frameworks becomes key, as improvement in governance, with a target of obtaining a better credit rating with reduced financing cost for renewables, is required. Another option will be to optimise the quality of the indigenous financial market as it enables the capacity of international agencies to increase the bulk of financial investment in renewable energy projects towards full electrification of the African continent.

References

Abanda, F.H., A. Ng'ombe, R. Keivani, and J.H.M. Tah. 2012. The Link Between Renewable Energy Production and Gross Domestic Product in Africa: A Comparative Study Between 1980 and 2008. *Renewable and Sustainable Energy Reviews* 16: 2147–2153.

Adeoti, O., B.A. Oyewole, and T.D. Adegboyega. 2001. Solar Photovoltaic-Based Home Electrification System for Rural Development in Nigeria: Domestic Load Assessment. *Renewable Energy* 24: 155–161.

AfDB. 2010. *Financing of Sustainable Energy Solutions*. African Development Bank Committee of Ten meeting, 1–16.

Afful-Koomson, T. 2015. The Green Climate Fund in Africa: What Should Be Different? *Climate and Development* 7 (4): 367–379.

Africa Progress Panel. 2015. *Africa Progress Report 2015*. Africa Progress Panel.

Barnes, D.F., H. Samad, and S.G. Banerjee. 2014. The Development Impact of Energy Access. In *Energy Poverty: Global Challenges and Local Solutions*, ed. A. Hal, B.K. Sovacool, and J. Rozhon. Oxford: Oxford University Press.

Benoit, P. 2006. *Energy for Africa, Sixth Meeting of GFSE, Africa Is energizing Itself*. Vienna.

Brew Hammond, A. 2010. Energy Access in Africa: Challenges Ahead. *Energy Policy* 38 (5): 2291–2301.

Briceno-Garmendia, C., K. Smits, and V. Foster 2008. Financing Public Infrastructure in Sub-Saharan Africa: Patterns and Emerging Issues. *Africa Infrastructure Country Diagnostic (AICD) Background Paper World Bank* 15.

Couture, T., and Y. Gagnon. 2010. An Analysis of Feed-in Tariff Remuneration Models: Implications for Renewable Energy Investment. *Energy Policy* 38: 955–965.

Daka, K.R., and J. Ballet. 2011. Children's Education and Home Electrification: A Case Study in North-western Madagascar. *Energy Policy* 39 (5): 2866–2874.

Dasappa, S. 2011. Potential of Biomass Energy for Electricity Generation in Sub-Saharan Africa. *Energy for Sustainable Development* 15: 203–213.

Deichmann, U., C. Meisner, S. Murray, and D. Wheeler. 2011. The Economics of Renewable Energy Expansion in Rural Sub-Saharan Africa. *Energy Policy* 39 (1): 215–227.

Department of Minerals and Energy. 2003. *White Paper on Renewable Energy; Republic of South Africa.* [Online]. Available from https://unfccc.int/files/meetings/seminar/application/pdf/sem_sup1_south_africa.pdf. 12 Feb 2018.

Duarte, M., S. Nagarajan, and Z. Brixiova. 2010. *Financing of Sustainable Energy Solutions: AfBD Committee of Ten Policy Brief.* Ivory Cost: Published by AfBD-African Development Bank.

Duku, M.H., S. Gu, and E.B. Hagan. 2011. A Comprehensive Review of Biomass Resources and Biofuels Potential in Ghana. *Renewable and Sustainable Energy Reviews* 15: 404–415.

Dunlop, S. 2008. *A Dictionary of Weather.* Oxford: Oxford University Press. ISBN: 9780191580055.

Eberhard, A., O. Rosnes, M. Shkaratan, and H. Vennemo. 2011. *Africa's Power Infrastructure: Investment, Integration, Efficiency.* Washington, DC: The World Bank.

Eggoh, J.C., C. Bangake, and C. Rault. 2011. Energy Consumption and Economic Growth Revisited in African Countries. *Energy Policy* 39 (11): 7408–7421.

Energy and Mines. 2016. *Renewable for Mines Driving Competitive, Secure and Low-Carbon Power Stations for Mines: Ministry of Energy, Mines, Water and Environment.* [Online]. Available from: http://www.mem.gov.ma/SiteAssets/Dicsours/Discours2016/London%20speech%20English.Pdf. 19 Jan 2018.

Eshun, M.E., and J. Amoako-Tuffour. 2016. A Review of the Trends in Ghana's Power Sector. *Energy Sustainability and Society* 6 (1): 1.

FAO (Food and Agriculture Organization of the United Nations). 2015. *FAOSTAT – Food and Agriculture Organisation of the United Nations, Statistics Division: Crops.* [Online]. Available from: http://faostat3.fao.org/browse/Q/QC/E. 6 Jan 2018.

Felix, M., and S.H. Gheewala. 2011. A Review of Biomass Energy Dependency in Tanzania. *Energy Procedia* 9: 338–343.

Geothermal Energy Association. 2015. *US-East Africa Geothermal Partner-ship (EAGP).* [Online]. Available from: http://www.geo-energy.org/EastAfrica/EAGP.aspx 13 Jan 2018.

Global Data. 2015. *Power Generation Technologies Capacities, Generation and Markets Database.* London: Global Data.

Global Alliance for Clean Cookstoves, 2014. *Cookstoves: The Issues.* [Online]. Available from: http://carbonfinanceforcookstoves.org/about-cook-stoves/introduction/. 9 Jan 2018.

Griffiths, S. 2017. A Review and Assessment of Energy Policy in the Middle East and North Africa Region. *Energy Policy* 102: 249–269.

Gujba, H., S. Thorne, Y. Mulugetta, K. Rai, and Y. Sokona. 2012. Financing Low Carbon Energy Access in Africa. *Energy Policy* 47: 71–78.

Gutman, J., A. Sy, and S. Chattopadhyay. 2015. *Financing African Infrastructure: Can the World Deliver?* Washington, DC: Global Economy and Development at Brookings Institution.

GWEC (Global Wind Energy Council). 2014. *Market Forecast for 2015–2019.* [Online]. Available from: http://www.gwec.net/global-figures/market-forecast-2012-2016/. 20 Jan 2018.

Heinrich Boell Foundation and World Future Council. 2013. *Powering Africa Through Feed-in Tariffs: Advancing Renewable Energy to Meet the Continent's Electricity Needs.*

Hochberg, M. 2016. *Renewable Energy Growth in Morocco: An Example for the Region.* [Online]. Available from: https://www.mei.edu/sites/default/files/publications/PF26_Hochberg_Moroccorenewables_web.pdf. 20 Jan 2018.

Hydropower and Dams. 2014. *Hydropower and Dams World Atlas.* Wallington: Aqua-Media International.

Ifelebuegu, A.O. 2013. Sustainability. In *Encyclopedia of Crisis Management*, ed. K.B. Penuel, M. Statler, and R. Hagen, vol. 1, 926–926. Thousand Oaks: SAGE Publications Ltd.

International Energy Agency (IEA). 2012. *Technology Roadmap: Hydropower: International Energy Agency.* [Online]. Available from: https://www.iea.org/publications/freepublications/publication/technology-roadmap-hydropower.html. 13 Jan 2018.

———. 2015. Africa Energy Outlook: A Focus on Energy Prospects in Sub-Sahara Africa. *World Energy Outlook; International Energy Agency,* 30–68.

———. 2016. *Policies and Measures: Kenya.* [Online]. Available from: https://www.iea.org/policiesandmeasures/renewableenergy/?country=Kenya 14 Jan 2018.

International Renewable Energy Agency (IRENA). 2013. *Road Transport: The Cost of Renewable Solutions.* Abu Dhabi: IRENA.

———. 2014. *Global Bioenergy Supply and Demand Projections: A Working Paper for RE Map 2030.* Abu Dhabi: IRENA.

International Renewable Energy Agency (IRENA) and IEA-ETSAP. 2015. *Hydropower: Technology Brief.* Abu Dhabi: IRENA.

International Renewable Energy Agency (IRENA). 2015a. *Renewable Energy Capacity Statistics 2015.* Abu Dhabi: IRENA.

———. 2015b. *Costs: Renewable Energy Costs, Technologies and Markets.* [Online]. Available from: http://costing.irena.org/. 10 Jan 2018.

———. 2015c. *Renewable Power Generation Costs in 2014*. Abu Dhabi: IRENA.

———. 2015d. *Renewable Energy Target Setting*. Abu Dhabi: IRENA.

IPCC. 2011. Special Report on Renewable Energy Sources and Climate Change Mitigation. In O. Edenhofer, R. Pichs-Madruga, Y. Sokona, K. Seyboth, P. Matschoss, S. Kadner, T. Zwickel, P. Eickemeier, G. Hansen, S. Schlömer, C.V. Stechow, ed. United Kingdom/New York/Cambridge: Cambridge University Press.

Karekezi, S., and W. Kithyoma. 2003. *Renewable Energy in Africa: Prospects and Limits*. [Online]. Available from: http://www.un.org/esa/sustdev/sdissues/energy/op/nepadkarekezi.pdf. 20 Jan 2018.

Kassenga, G.R. 1997. Promotion of Renewable Energy Technologies in Tanzania. *Resources, Conservation and Recycling* 19 (4): 257–263.

Kammen, D.M. 2015. Peace Through Grids. *MIT Technology Review*, May/June.

Kolawole, A., S. Adesola, and G. De Vita. 2017. A Disaggregated Analysis of Energy Demand in Sub-Saharan Africa. *International Journal of Energy Economics and Policy* 7 (2).

Lerner, A.M. 2010. *SADC Biofuels State of Play Study*. Gaborone: SADC Secretariat Energy Division.

Linklaters. 2016. *Renewable Energy in Africa: Trending Rapidly Towards Cost – Competitiveness with Fossil Fuels*. [Online]. Available from: https://www.linklaters.com/en/insights/thought-leadership/renewables-africa/renewables-in-africa. 24 Jan 2018.

Martinot, E., A. Chaurey, D. Lew, J.R. Moreira, and N. Wamukonya. 2002. Renewable Energy Markets in Developing Countries. *Annual Review of Energy and the Environment* 27: 309–348.

Ministry of Energy, Mines, Water and Environment. 2015a. *Draft Law No. 48–15 Relating to the Regulation of the Electricity Sector*. [Online]. Available from: http://www.mem.gov.ma/SitePages/GrandChantiersEn/DERegulationOfSector.aspx#. 2 Feb 2018.

———. 2015b. *Achievements for Strengthening Electric Production and Transport Capacity (2009–2015)*. [Online]. Available from: http://www.mem.gov.ma/SitePages/GrandChantiersEn/DEProductionTransport.aspx. 20 Jan 2018.

Mushtaq, S., T.N. Maraseni, J. Maroulis, and M. Hafeez. 2009. Energy and Water Trades in Enhancing Food Security: A Selective International Assessment. *Energy Policy* 37 (9): 3635–3644.

Ondraczek, Janosch, Nadejda Komendantova, and Anthony Patt. 2015. WACC the Dog: The Effect of Financing Costs on the Levelized Cost of Solar PV Power. *Renewable Energy* 75: 888–898.

Pegels, A. 2010. Renewable Energy in South Africa: Potentials, Barriers and Options for Support. *Energy Policy* 38 (9): 4945–4954.

REN 21 (Renewable Energy Policy Network). 2016. *Renewables Global Status Report.* [Online]. Available from: http://www.ren21.net/wp-content/uploads/2016/06/GSR_2016_Full_Report.pdf. 15 Jan 2018.

Rickerson, W., C. Hanley, C. Laurent, and C. Greacen. 2013. Implementing a Global Fund for Feed-in Tariffs in Developing Countries: A Case Study of Tanzania. *Renewable Energy* 49: 29–32.

Simiyu, S.M., and G.R. Keller. 2000. Seismic Monitoring of the Olkaria Geothermal Area, Kenya Rift Valley. *Journal of Volcanology and Geothermal Research* 95: 197–208.

UNEP. 2012. *Financing Renewable Energy in Developing Countries.* United Nations Environment Programme Finance Initiative.

UN-HABITAT (United Nations Human Settlements Programme). 1993. *Application of Biomass Energy Technologies.* Nairobi: UN-HABITAT.

United Nations (UN). 2015. *Transforming Our World: The 2030 Agenda for Sustainable Development Doc. A/70/L.*

———. 2018. *Report of the World Commission on Environment and Development.* [Online]. Available from: http://www.un.org/documents/ga/res/42/ares42-187.htm. 16 Jan 2018.

United Nations Commission on Sustainable Development. 2015. *South Africa Country Report.* [Online]. Available from http://www.un.org/esa/agenda21/natlinfo/countr/safrica/energy.pdf. 13 Jan 2018.

US DOE. 2017. *Annual Energy Outlook.*

UNFCCC. 1997. *The Kyoto Protocol, Article 12. In: Third Conference of the Parties to the UNFCCC.* Kyoto.

Winkler, H., A.F. Simões, E.L. La Rovere, M. Alam, A. Rahman, and S. Mwakasonda. 2011. Access and Affordability of Electricity in Developing Countries. *World Development* 39 (6): 1037–1050.

World Bank. 2014. *From the Bottom Up: How Small Power Producers and Mini-Grids Can Deliver Electrification and Renewable Energy in Africa,* Washington DC: The World Bank. World Bank [Online]. Available from: https://data.worldbank.org/indicator/EG.ELC.ACCS.ZS

———. 2017. *Regulatory Indicators for Sustainable Energy.* Available from: http://rise.worldbank.org/. Accessed 1 Mar 2018.

World Trade Organisation. 2007. *Trade Policy Review Report by the Secretariat WT/TPR/S/194.* [Online]. Available from: http://www.wto.org/english/tratop_e/tpr_e/s194-00_e.doc. 17 Jan 2018.

9

Conclusions and Emerging Perspectives on Africa's Energy: Can Multiple Perspectives on Energy Management Inform Policy and Practice?

Sola Adesola and Feargal Brennan

This book promotes dialogue and new thinking in the broad field of energy, regional development, sustainability, and international business management. Contributions have to an extent questioned and debated on the hegemony of policy, country differences, managerial orthodoxy, and the dominant academic thinking on energy research but perhaps limited discourse on bridging the gap between science and social science on energy. Future work could cover conceptual, theoretical, and empirical work that debates issues linking energy research and management and international business research that utilise different theoretical perspectives and apply a wide variety of rigorous methodological approaches.

S. Adesola (✉)
Oxford Brookes Business School, Oxford Brookes University, Oxford, UK
e-mail: sadesola@brookes.ac.uk

F. Brennan
University of Strathclyde, Glasgow, UK
e-mail: feargal.brennan@strath.ac.uk

© The Author(s) 2019
S. Adesola, F. Brennan (eds.), *Energy in Africa*,
https://doi.org/10.1007/978-3-319-91301-8_9

This book would be complete to understand the need to foster collaborations among academia, industry, and government as well as evaluating the country/regional participation in clean energy technology development. Future research could explore the transfer of clean energy technologies between the government, universities, and the industries.

Future Trends and Developments Shaping the Continent

In this concluding chapter, an attempt is made to identify common and emerging issues of energy across African countries as well as highlighting industry and country specific issues. Understanding the directions in which Africa's energy sector is set to develop is essential for policymakers and industry practitioners.

Policy Development

A number of questions are being posed from policy that shape and influence the direction of energy. This book therefore raises the following questions and issues for further reflection and study:

- What is the role and influence of national, regional, and international institutions in shaping energy markets? Policy development and co-ordination at continental and regional level are undertaken by the African Union (AU) and the New Partnership for Africa's Development (NEPAD), which have formulated the AU/NEPAD African Action Plan, and, with the African Development Bank, the Programme for Infrastructure Development in Africa (PIDA). Much of the policy focus at this level is on trans-national infrastructure development. A number of multilateral and bilateral initiatives interplay with national plans, such as the US Power Africa initiative; Sustainable Energy for All Initiative; Energising Development initiative (European Union); Energy+ (Norway, UK, and others); and EnDev programme (Germany, Norway, and others). An example of policy co-operation is the

Africa-EU Energy Partnership 2020 targets and its related programme to develop renewable energy markets (the Africa-EU Renewable Energy Co-operation Programme). In addition, there is a broad range of civil society-led initiatives that are often in line with national energy objectives while not necessarily linked to them explicitly.

- Regional energy integration poses huge potential for combining approaches to data, finance, and renewable. Future work might want to also consider the legal and policy regulations on international trade laws in energy and the domestic energy policy dynamics.
- Role of the regulator in ensuring sustainability of the gains of privatisation in a developing country.
- Institutional and regulatory dimensions of energy security, though discussed in the context of international gas supply market, are broadly applicable as a critical issue for further study.
- Investment criteria for energy financiers. If improvements in the energy policy and regulatory framework are successful, the next hurdle relates to sources of capital to fund energy projects.

The level of active commitment by government and other stakeholders is considered, as is the extent to which regulatory and financing issues have been resolved. Securing a more prosperous future for Sub-Saharan Africa (SSA) depends on progress in three areas of energy policy: increased investment in supply, in particular of electricity, to meet the region's growing energy needs; improved management of natural resources and associated revenues; and deeper regional co-operation. Regional co-operation is a major element of Africa's vision for its future, providing a cost-effective way to increase the availability and security of energy supply. The pace of change will be set by the quality and integrity of the public institutions concerned, as well as the transparency and accountability of their operations.

Management

Understanding international energy market—despite the need for understanding energy business as a complex global phenomenon, there is a general lack of research addressing environmental challenges from an international business management perspective. The extant literature on energy is mostly studied from technical and country boundaries without interest in studying the interdisciplinary and cross-border dynamics. This edited book focuses on the development of current status and policies; a few studies have been published on how regional and international practices or countries cross-border institutional development and geopolitical influence the investment, competitiveness, and multinational enterprise (MNE) interaction in the era of energy transition (Kolk 2015, 2016).

Given the energy trend of MNEs transitioning towards renewable energy sources (Hardy 2016), renewable energy has the potential to be a fascinating topic to study for international business and management scholars. Africa being blessed with renewable energy sources, solar photovoltaics is scalable and surplus in Africa (Kaartemo 2016); however, resources and demand for solar and other renewable energy investments are regional and global. This book has seen the interest and influence by international institutions and agencies on energy sources; for example, the Paris Climate Agreement was signed by 194 nations in 2015, yet these agreements have not been implemented locally in Africa due to different capacities and institutional regulatory environments for attracting and discouraging investors, and driving environmental innovations (Gonzalez-Perez 2016). At the same time, Africa does not have the same degree of costs associated with moving to renewable and clean energy infrastructure as in Organisation for Economic Co-operation and Development (OECD) countries for example. The advantage of having a "blank canvas" to design and develop modern, sustainable energy infrastructure optimised for local demand can be underestimated.

Future research on energy in Africa can open up new perspectives on regional integration, impact of funding and investors to generate social or environmental impact alongside a financial return, and the issue of affordability. Research looking at investment enablers and disablers of renewable energy and innovation would further expand our knowledge and practice. Furthermore, being heavily influenced by policy and sustainability, management and international business research could provide new insights on the regional developments of energy and renewable energy markets and provide international perspective on entrepreneurship on renewable energy. The chapter on innovation adoption theory in the context of solar technology in Uganda has created a space for ongoing conversation on innovation and disruption in the energy sector, reflecting the rapid change and high rates of innovation. It would be interesting in the future to consider how energy economics and policy analysis can help chart a course through the uncertainties, facilitating technology development, investment and securing consumer trust towards secure and affordable sustainable energy system in the long term. In the words of Dr Elham Ibrahim, Regional Vice Chair for Africa, World Energy Council, at the 10th Annual Africa Energy regional meeting in February 2018:

> *Innovation and collaboration are key drivers for the grand energy transition. Africa has forged close ties with many countries and regional and multilateral entities to further develop collaborations. The goal is to identify opportunities to further promote innovation.* (World Energy Council 2018)

We further focus on two areas that are critical to a better-performing SSA energy sector: better management of the region's resources and deeper regional energy co-operation. These conditions are inter-linked to the standards of governance for success. Future research could investigate governance failure in the energy sector and investment in the skills and knowledge required for a modern energy economy and the transparency and consultation on energy policies that is essential to winning public consent.

Better Management of the Region's Resources SSA has ample energy resources, both fossil fuel and renewable, but the opportunities that it offers to support sustained economic growth are often missed. A glaring example is the way that deficiencies in essential infrastructure in many countries are perpetuated by ineffective or corrupt misuse of revenues from fossil fuel extraction. The example of Botswana suggests that the risks associated with high dependence on a single resource can be mitigated. Botswana shows a model for resource governance, and its success rests on several interwoven factors. The country has maintained a multiparty democratic political system since independence, with an established culture of accountability and transparency that is anchored in the Tswana traditions of consultations, participation, and consensus. Consistently rated amongst the least corrupt African countries, Botswana planned its public spending over time, and any domestic infrastructure investments are ratified by parliaments.

Deeper Regional Energy Co-operation and Integration Expanding cross-border trade can be a very cost-effective way to increase the reliability and affordability of energy supply, but this is often hindered in practice by a range of technical and political barriers. The lack of regional scale is a particular obstacle for the development of SSA's large remaining hydropower potential. Given that regional co-operation is a major component of Africa's vision for its future, further study on integrated regional power markets would be essential to understand the potential gains from regional co-operation and the actual record of achievement. Research could be conducted on successful cross-border co-operation and cross-border infrastructure in SSA—for instance, the gas pipelines from Mozambique to South Africa and the West Africa Gas Pipeline.

Given the inter-disciplinarity of this book, the issue of language becomes important. Engagements with energy experts raise the pedagogic question of how non-energy scholars can better understand energy experts.

Improving the relatively poor state of the existing infrastructure as a contribution towards a more modern energy system will require a much expanded skilled and semi-skilled workforce throughout the energy sector, including technical skills, as well as skills related to policy, regulation, and project management. Skill levels in Africa are collectively increasing with each project, based on localisation requirements often built into the procurement process. The need to invest in building human capacity is increasingly recognised and is reflected in projects such as the EU Energy Initiative—Partnership Dialogue Facility (EUEI PDF) and Barefoot College, which train solar engineers in rural communities. As the population of SSA receives less than 5% of schooling on average (UNDP 2013), this suggests that the level of education and skills will remain a key challenge.

Sustainability

The energy projection has a wide range of environmental and societal implications, both in terms of local impacts in Africa and the much broader issue of global climate change. The linkage between energy research and management can further advance our understanding of the influence of corporate sustainability and corporate social responsibility adoption of clean energy and the thematic area of renewable energy. The diverse issues are important factors for policymakers to monitor and tackle and often require ongoing social and corporate engagement with a number of stakeholders of the use of land by energy MNEs. The Niger Delta contaminated land case study provides huge opportunity for other countries to develop their energy resources in a sustainable manner. This is an opportunity for research development in the future. The issue of affordability and cost of technologies are vital questions worthy of further analysis if energy in Africa is to achieve sustainability.

Finally, it is expected that the knowledge and insights provided in this compendium would enable academics, industry leaders, government, and institutions to make more informed choices about the world's most pressing energy challenges and opportunities in Africa.

References

Gonzalez-Perez, M.A. 2016. Climate Change and the 2030 Corporate Agenda for Sustainable Development. *Advances in Sustainability and Environmental Justice* 19: 1–6.

Hardy, Q. 2016. Google Says It Will Run Entirely on Renewable Energy in 2017. *The New York Times*, Available at: http://www.nytimes.com/2016/12/06/technology/google-says-it-will-run-entirely-on-renewable-energy-in-2017.html?_r=0.

Kaartemo, V. 2016. Creation and Shaping of the Global Solar Photovoltaic (PV) Market. *Advances in Sustainability and Environmental Justice* 19: 229–250.

Kolk, A. 2015. The Role of International Business in Clean Technology Transfer and Development. *Climate Policy* 15 (1): 170–176.

————. 2016. The Social Responsibility of International Business: From Ethics and the Environment to CSR and Sustainable Development. *Journal of World Business* 51 (1): 23–34.

UNDP (United Nations Development Programme). 2013. *Human Development Report 2013 the Rise of the South: Human Progress in a Diverse World.* New York: UNDP.

World Energy Council. 2018. Available from https://www.worldenergy.org/news-and-media/news/solutions-for-africa-at-the-10th-energy-indaba/.

Correction to: Chapter 1: Introduction to Energy in Africa: Policy, Management, and Sustainability

Sola Adesola and Feargal Brennan

Correction to:

Energy in Africa

https://doi.org/10.1007/978-3-319-91301-8

This chapter was inadvertently published with errors. Footnotes, citations and references were added.

The updated online version of this chapter can be found at
https://doi.org/10.1007/978-3-319-91301-8_1

S. Adesola (✉)
Oxford Brookes Business School, Oxford Brookes University, Oxford, UK
e-mail: sadesola@brookes.ac.uk

F. Brennan
University of Strathclyde, Glasgow, UK
e-mail: feargal.brennan@strath.ac.uk

© The Author(s) 2019
S. Adesola, F. Brennan (eds.), *Energy in Africa*,
https://doi.org/10.1007/978-3-319-91301-8_10

Index[1]

---

[1] Note: Page numbers followed by 'n' refer to notes.

Printed by Printforce, the Netherlands